In Search of Ancient Tsunamis

In Search of Ancient Tsunamis

A Researcher's Travels, Tools, and Techniques

JAMES GOFF

OXFORD
UNIVERSITY PRESS

OXFORD
UNIVERSITY PRESS

Oxford University Press is a department of the University of Oxford. It furthers the University's objective of excellence in research, scholarship, and education by publishing worldwide. Oxford is a registered trade mark of Oxford University Press in the UK and certain other countries.

Published in the United States of America by Oxford University Press
198 Madison Avenue, New York, NY 10016, United States of America.

Library of Congress Cataloging-in-Publication Data
Names: Goff, James R. (James Rodney), 1959– author.
Title: In search of ancient tsunamis : a researcher's travels, tools, and techniques / James Goff.
Description: New York : Oxford University Press, [2023] |
Includes bibliographical references and index.
Identifiers: LCCN 2022040716 (print) | LCCN 2022040717 (ebook) |
ISBN 9780197675984 (hardback) | ISBN 9780197676004 | ISBN 9780197675991 (epub) |
ISBN 9780197676011 (ebook)
Subjects: LCSH: Tsunamis—Research. | Natural disasters—History.
Classification: LCC GC221.2 .G64 2023 (print) | LCC GC221.2 (ebook) |
DDC 551.46/37—dc23/eng/20221012
LC record available at https://lccn.loc.gov/2022040716
LC ebook record available at https://lccn.loc.gov/2022040717

DOI: 10.1093/oso/9780197675984.001.0001

1 3 5 7 9 8 6 4 2

Printed by Sheridan Books, Inc., United States of America

To Genevieve—my inspiration and recipient of a seemingly endless stream of words and thoughts, some of which I hope are worth listening to.

Contents

Illustrations

Figures

Plates

Preface

We have thus a theory of those waves which seem almost invariably to have accompanied the great earthquakes; supposing these to have been occasioned by submarine elevations.

J. Hall, On the Revolutions of the Earth's Surface

I can almost hear someone saying, "Do we really need yet another book about tsunamis?" The answer to that is quite simple. Yes, we do. We actually need many more books about tsunamis, if only for education and awareness.

This book is slightly different to the normal tsunami book, if there is such a thing. It is really a story about how we go about finding evidence for past tsunamis. How do scientists identify these things? During this finding out process in the book we learn not only about some of the techniques scientists use but also about the events themselves, and in the telling we learn a bit about the person who does the work and the places that work is done in, and we dip our toes into the odd anecdote or two about the trials and tribulations of a tsunami researcher. So yes, it is educational, but it is told from the perspective of a researcher essentially charting his progress through this new field of research as it has developed. I learn, you learn.

Charting my pathway through this developing field of tsunami research was not done in isolation. The strength of this area of research lies in the fact that it is not the preserve of any single discipline, far from it, the very nature of this beast requires a multidisciplinary approach to fully get to grips with it. This may come as a surprise to the geologist who considers that sediments and deposits are all there is to it, or to the mathematical modeler who can predict or postdict a tsunami moving through the water column. Ultimately, while it is essentially to understand the nature of each individual tsunami by studying the evidence it leaves behind and by tying this together through models, the main reason we do this is because tsunamis kill, destroy, and wreak havoc along the world's coastlines. In other words, there is a human side to this story, where humans and the environment interact in a brutally one-sided contest. This interaction has been going on for millennia and in many ways tsunamis have challenged our ability to ensure a long-term and stable relationship with the coast.

Tsunamis continue to surprise us—the 2004 Indian Ocean earthquake tsunami is a case in point, as was the more recent 2022 Hunga Tonga-Hunga Ha'apai eruption and tsunami in Tonga. We may feel cossetted by 21st century technology, but it has manifestly failed to protect our lives and livelihoods from nature's terrorists. That is why we delve into the past to better understand tsunamis. As we have dug deeper, it has become patently obvious that far bigger and nastier events are waiting for us in our future, because they have happened in the past and will happen again. That is why the study of tsunamis is also as much the domain of the archaeologist and the anthropologist. Our ancestors died, their communities were ripped apart, some survived, they recovered, adapted, became more resilient, relaxed, forgot, and the cycle repeated itself, and it continues to repeat itself to this day. While we continue to repeat mistake after mistake, it is also through the repetition, again and again, of a coherent and persistent educational message that we may break this cycle and find a way to live in a form of punctuated equilibrium with one of nature's unpredictable eccentricities. It is only through a better understanding of tsunamis that we can learn to adapt. The desperate human drive to populate and use the coast must be tempered by a respect for the sea. Will we ever attain such a balance? As a scientist I strive to contribute to this lofty goal if for no other reason than that the alternative means more deaths, more soul searching, more mistakes, repeated time after time after time.

In telling this very personal story I have dipped into my mental and written archives. Like all such records they are manifestly incomplete, and in many cases the memories have blurred over time. Some memories are as clear and bright as the day they happened . . . or are they? Ask several people to describe an event that happened a matter of minutes ago and you will get a different story every time. So my first apology is to colleagues and acquaintances, who while not named will recognize the kernel of an experience that they thought they experienced. It undoubtedly was that experience that my anecdote refers to, but it is seen not only through my eyes, but in some cases deliberately merged with another one, or concertinaed in time, or embellished somewhat simply because it fits better with the flow of the story. My second apology, which isn't really an apology, is to what one hopes will be a wide variety of readers. This is not a textbook. It is not a dry, dusty, and predictably methodical tome through which a student is forced to plod in

order to get certain information fed into their brains. It is hopefully an enter-
taining, educational, and page-turning read that will end too soon and leave
the reader wanting to learn more. My final apology is to those readers who do
go from start to finish. The end is not the end. As I type, we have made new
and exciting discoveries related to what I mention in the final chapter but as
scientists we want to do a little more work before unleashing our excitement
on the world. Suffice it to say, there will be another book, maybe not solely
about tsunamis, but rather what our work on tsunamis adds to our understanding of what life and death has meant to coastal communities over the
millennia, and what this means for our future at a time when sea level rise is
on steroids, geologically at least.

Hopefully you will enjoy this journey of discovery, I have.

Acknowledgments

This book would never have seen the light of day if I had not had the pleasure of working with so many wonderful colleagues over the years. It is impossible to name all of you for which I apologize, but there are a few in particular who must get an honorable mention.

There have been innumerable fieldtrips, all of which have been memorable in some way, shape, or form. On many of these I was lucky enough to be accompanied by Bruce McFadgen whose sense of humor, fierce intellect, and impossibly fast walking made every day so enjoyable. Equally, the remarkable Scott Nichol managed to make so many of our trips an unforgettable experience. Entire books could be written about the fun of working with these two gentlemen—thank you. James Terry, Walter Dudley, Bruce Jaffe, the late Bruce Richmond, Kazuhisa Goto, Daisuke Sugawara, Gabriel Vargas, Diego Salazar, Pedro Andrade, Geoffroy Lamarche, Fred Taylor, and many more are thanked for so many brilliant adventures. Particular thanks for to Bruce McFadgen, Scott Nichol, Pedro Andrade, Andrew Wells, and Fred Taylor for the use of images that were important for this book. There have been many people that I have met on this journey who have both inspired me and helped me in so many ways. To all of you I say thanks. Thank you for helping me try to unravel the mysteries of these killer waves. I hope that we have many more miles on the clock yet because the journey is far from complete.

I thank my family for their endless support and interest in their brother's/ uncle's/son-in law's/brother-in-law's odd vocational activities. Finally, I could not have done this without the support, love, and encouragement of Genevieve Cain, who has hung in there with me throughout the entire process, again.

1

Serendipity

An Introduction

Subject: Invitation from Chile
Date: Saturday, 24 September 2016 5:05:54 PM AEST

This is the story of a journey. It begins in a small, slightly overfull office in Sydney, Australia, and ends in the unremitting heat of the Atacama Desert coast in Chile. To get from A to B, though, is no straight line, it takes many geographical and anecdotal tangents. The journey is one of discovery. It describes my part in the emergence of a new and extremely violent science—unravelling the mysteries of past tsunamis.

There are two sides to the story of every tsunami—the careful piecing together of the physical evidence for these catastrophic events and the patient assembly of the jigsaw puzzle that often reflects the less tangible signs, the human story. After all, the only reason we really care about tsunamis is because they catch us by surprise, destroy our coastal communities, and kill us. The past couple of decades have seen thousands of unnecessary tsunami-related deaths but we continue to believe that we can control nature, control this beast. It is only when you delve into the past that you start to realize that not only has the world experienced bigger tsunamis than those in living memory, they have also destroyed coastal communities, and just like today we never seem to have truly learned from these experiences. Despite this huge back catalogue of human folly, we continue to make the same mistakes. In this case the past is our key to the future—the more we know about the past, the better prepared we will be for the future. The trick is to know what you are looking for, and this is the start of that journey.

In 2016 I resigned as professor of natural hazards with the intention of spending more of my life in the field and writing up an outstanding pile of

In Search of Ancient Tsunamis. James Goff, Oxford University Press. © Oxford University Press 2023.
DOI: 10.1093/oso/9780197675984.003.0001

past research. And so I entered semi-retirement (do academics ever fully retire?) with no regular salary, no immediate research on the horizon, but always the hope that something would materialize.

This was a perfect scenario for serendipity—free of the shackles of an academic job, footloose and fancy free, but with a brain that would doubtless go into meltdown without something to do (not a very well-planned semi-retirement!). It was crying out for serendipity to strike. At the time the serendipity of this particular moment was not obvious, it is only through the luxury of hindsight that this moment stands out so clearly.

In science, geology in this case or at least it is to start with, serendipitous events are more likely to occur in a recently born subject where each new discovery, each new strand of carefully researched material, stretches out into the unexplored void that is waiting to be filled. This may all sound a bit dramatic, perhaps it's better to say that it is a bit like when you happen to hear or read a new word and after finding out what it means you suddenly start to notice it more and more often. Of course the truth is that the word had always been there but because of some weird mental blind spot it drifted through the brain rather than registering in it.

Each strand of this new material, then, is thrust out into the scientific melting pot, often as a peer-reviewed paper invariably doomed to die a death and never see the light of day again. If there was a library of forgotten science a bit like Carlos Ruiz Zafón's Cemetery of Forgotten Books in his novel *The Shadow of the Winds*, it would rival the Library of Congress for the sheer volume of sometimes brilliant ideas that languish in obscurity. But every now and again one of these forgotten strands meets another and an idea is born.

A case in point.

University libraries can be fun, or at the very least they are repositories of fascinating pieces of information, most of which you do not know you need or want until you find them. So basically they are worth visiting from time to time. There is no substitute for the tactile experience of a real book, and then there is the smell of them and the pleasures to be gained by reading the words from the physical pages and not some computer-generated Portable Document File or PDF that we have become so familiar with. An author I know once insisted that her book be printed on hand-made paper—to hell with the expense, to hell with profits, bring back the tactile experience was her battle cry. And it worked, and there were profits too. There are enough people out there desperate for that "book" experience to ensure that an author's eccentricities are not as eccentric as we would like to think.

It is in the dry, dusty, forgotten tomes that linger menacingly in those rarely visited sections of the library, threatening to crush the unsuspecting browser, that gems can be found. You know when you have arrived in one of those sections because the smell of the slowly decaying books is pervasive. There are few libraries capable of providing the wherewithal to halt this decay, and in many ways this is to be celebrated. Libraries would lose a lot of their mystique without that smell of old books. Sadly though, these impressively imposing volumes of long forgotten journals are slowly being excised from the shelves with scalpel-like precision by administrators desperate for space, or simply out of malice. On occasion a student might wander down these weathered, careworn aisles, looking aimlessly at one title after another, but rarely do they stop, savor, and explore the writings of our forebears.

I however, had a mission. There was an old paper that possibly had the germ of an idea related to an ongoing project of mine on tsunamis. Having deliberately searched out this paper in a not entirely forgotten old journal, it was something of a surprise to find another small gathering of journal volumes that I never even knew existed right next to it in the stacks. These were the volumes of *The Early Settlers and Historical Association of Wellington*.

These beautifully bound volumes were almost pristine, untouched by years of enquiring fingers, untainted by the contaminating traces of fast food, coffee, and life in the 21st, or even 20th, century. These were clearly forgotten, parked up into a side alley of academe, awaiting their inevitable fate that would one day see them unceremoniously dumped into a skip sitting on the ground below a window in their fifth-floor hideaway. I had seen this happen before, and so managed to acquire some wonderful 19th century journal volumes surplus to the requirements of a modern library. This wanton disposal of academic musings is invariably justified by pointing out that hard copies still exist elsewhere and anyway they are available as PDFs for those adept at internet searches. Gone are the fortuitous finds, the idle leafing through old volumes, the title that springs off the page and rings an alarm bell or possibly turns on a light bulb in the back of your mind. The assumption is that these fortuitous finds will occur on the web, your searches will inadvertently draw these out from the ether. Well, yes and no. Internet searches come with their own fortuitous rabbit holes, but these can never replace that wonderful feeling of discovery as you leaf through an old volume, cosseted in the dreamy miasma of old book smell, touching the physical pages of the past, and realizing once again that there is nothing new under the sun, your grail had just been mislaid in the forgotten library of academic endeavor.

It transpired that the old tomes I had just stumbled upon represented the works of one of New Zealand's oldest historical societies. The Early Settlers and Historical Association of Wellington was formally incorporated in June 1912 when there were still a number of Wellingtonians alive whose childhood's dated back to the 1840s and 1850s when the area was first settled by Europeans. The associations felt, quite wisely, that it was important that the stories of these aged Wellingtonians be recorded before they were lost, in order to capture the essence of one of the world's last colonial experiences. The first issue of the first volume was published in 1912 but the intervention of World War I and its aftermath meant that the first issue of the second volume was not printed until June 1922. Three further issues were printed, but in May 1923 the association concluded that this journal, or magazine as they called it, was not financially viable and publication ceased. There were not many volumes then, and they were sandwiched between far more prestigious tomes, doomed to be scrapped as soon as some suitably myopic library administrator could find them no doubt.

The immediate thought that came to mind when stumbling across these gems was what of interest, if anything, was hidden in their pages.

It didn't take long to find it.

In Volume 1, from 1913, Issue 3, on pages 114–117, was the article "Pre-Pākehā Occupation of Wellington District" by Hector Norman McLeod, one of the original committee members of the association. True, this was not a very auspicious title (Pākehā essentially means a white New Zealander as opposed to a Maōri). BUT experience has tended to show that not a lot of thought went into capturing the reader's attention with a title in those days, it was all about what was in the paper—you had to read it to find out.

On page 116 in among a general discussion about a ramble around the area fossicking for bits and pieces, McLeod mentions almost as an aside, "a number of cetacean skeletons, crumbling to powder, yet preserved in form in dry sand, lie at heights up to 147 ft above high water mark. One, stretching over 60ft, is half a mile inland."

What?

A lovely old map of the Wellington District included in the paper highlighted all the finds mentioned in his paper (Figure 1.1), including the whale skeletons (Point 19 on the map), along the south coast of Miramar Peninsula. This peninsula lies immediately to the east of Wellington International Airport on the northern edge of Cook Strait, although there was no airport back then.

Figure 1.1 Hector Norman McLeod's map of Miramar Peninsula showing the site of the whale remains (19—enclosed by grey circle) in the southern part of the peninsula adjacent to Cook Strait. (From McLeod, 1913)

A few months later I had time to visit the area. A quick hop, step, and a jump—train, plane, and automobile—and I was there, standing on top of the cliffs overlooking the wild intimidating roughness of the seas of Cook Strait. This ridiculously implausible rabbit hole of a sentence had definitely been worth exploring "on the ground," because I had found another reference to the whales in a newspaper article. The November 3, 1908, issue of the Wellington *Evening Post*, threw some interesting light on the matter. Clearly, Mr McLeod had reported his cetacean skeletons some years before he wrote his paper. In the article the reporter was pondering how they got there, "If the whales were washed there then one of two hypotheses may be accepted: that a tidal wave carried the carcasses far inshore, or that the land has been raised with the embedded carcasses in it."

A tidal wave???

Standing up there on the cliff put a lot of things in perspective. Cook Strait has long been recognized as a major sea route for the sperm whale (*Physeter macrocephalus L.*), a shortcut between the Tasman Sea and the Pacific Ocean. Not only that, but sperm whales tend to travel in groups ("a number of cetacean skeletons") and the average size of the male (called a bull) is about 16 m (52 ft). The biggest that McLeod found was 60 ft long so that was on the money. Given the context of the find, a staggering 147 ft above the raging seas of Cook Strait, it was knee tremblingly scary to find out that a mature sperm whale weighs in at around 41 tonnes (45 tons).

The knee trembling may sound slightly dramatic until you realize that this was serendipity in action. It was 2009, and my finding coincided with new research that had just identified several large submarine landslides in Cook Strait; landslides that would have generated tsunamis!

There must have only been the most infinitesimally small likelihood that a pod of whales was passing through Cook Strait at the same moment a submarine landslide occurred and generated a tsunami, but it happened. Standing there on top of the cliff and looking down at the sea truly brought home to me the power of that tsunami. But to those unfamiliar with the location, a question will linger like the proverbial elephant in the room. If this was indeed the result of a (very large) tsunami, then why had no one ever found evidence for this event in the hundred years since McLeod made his observations? The answer to that is sadly very simple. The whales and the sand they lie in, quite possibly the only obvious evidence for this event, have been built on. There is now a beautiful, cliff-top subdivision of gorgeous modern houses with stunning views . . . perhaps of the next tsunami.

Serendipity—the strands from 1908 and 1913 met the strand from 2009. What on earth were the odds of a tsunami researcher being in the right place at the right time, being side-tracked by the appearance of a few obscure journal volumes in his peripheral vision, and then reading the long-forgotten pearls of wisdom that breathed new life into modern science?

With serendipity comes clarity, when these strands touched they started to inexorably bind themselves together, and when the sheer logic of such linkages comes to light, a veritable eureka moment ensues. The strand that had been hanging in a void, unsupported by any other data, suddenly has a direction and a purpose. The resulting paper needed to have a title that somehow gave a nod to serendipity and the time period in which the whale skeletons were discovered. Invoking the sage words of the fictional private detective Sherlock Holmes, the title of the 2009 paper was "Cetaceans and tsunamis—whatever remains, however improbable, must be the truth?"

In the study of ancient tsunamis, and that is where we are heading on this journey through the rabbit holes of science, such moments are actually more common than one might think.

Tsunamis are not new

In the recent past, tsunamis have become big news. The modern world received a wake-up call from a devastating event in Indonesia in 2004, and just as we were relaxing after that one, another brutal wave struck Japan in 2011. Tsunamis are not new though, although tsunami research, tsunamology, or whatever anyone wishes to call it, is still finding its way, attempting to carve out its own niche in the hazard world, sandwiched between the well-established and well-funded big boys of earthquakes and volcanoes. However, this is not a story about the early tsunami work done by a handful of talented scientists, although that is indeed worthy of a book, nor is it a once-over-lightly of everything that you probably never wanted to know about tsunamis. This book is a story about sudden, exciting discoveries, that make this new kid on the block a force to reckoned with and have taken the study of tsunamis from a little league into a science that can bring some serious findings to the table, enriching our understanding of how humans lived and died many thousands of years ago—and offering us a window into the future.

Ultimately, this is a tale about what has been hiding in plain sight for many years, but that could only be revealed once all the necessary building blocks were in place. And yes, in the absence of serendipity, we would undoubtedly not be where we are today.

It is all very well having an end point, or at least a nexus of strands in a continuum of discovery, but we need to try to define a beginning, a first strand that starts the process of discovery. It was not the whales on the cliff, although they are a wonderful example of exactly the type of thing that can come out of left field, wave its arms in the air, and shout, "look at me, look at me." So where does the first strand in the story begin? The answer seemed obvious to me before I started writing the next sentence but then as soon as I started, my mind ran ahead of me, and a list of aspiring first strands started to appear.

The best starting point for this story is probably in New Zealand (that was a great start, but the mind kept going), or perhaps Papua New Guinea, or even the Solomon Islands, or Chile—it is definitely in the Southern Hemisphere, although in saying that we must not forget Greece, well the Mediterranean really, or perhaps the North Sea or the United States.

And there we have it, even for a young science, it is difficult to put your finger on precisely where the first strand should come from, in fact it quickly becomes obvious that there were many "first" strands depending upon which piece of the puzzle you favor. Picking any one of these as the first strand immediately biases the entire story.

Let's stop dancing around this subject though and start getting serious.

The study of tsunamis is important.

It is important because the more we learn about tsunamis the better prepared we can all be for when the next big one strikes, and please note that is WHEN, not IF. But how BIG is BIG? We care about this because tsunamis kill people. The big ones, or at least the ones that we think are big, such as the 2004 Indian Ocean tsunami that killed well over 200,000 people, catch everybody, including scientists, off guard. But the surprise caused by the 2004 tsunami just goes to show how little we knew at the time, and how complacent we were about this killer. It is not like this was the first tsunami that killed people. But just how bad was the 2004 event in the big picture of the history of these destructive waves?

To understand tsunamis we need to realize that there are actually TWO ways of thinking about them. As humans we tend to focus on them as a hazard, capable of causing harm to people, to our communities, to our livelihoods, and to the landscapes that we love. But they are also quite simply a natural process that occurs within the environment we live in. They have been going on far longer than we have been around and will doubtless continue long after we have gone. But for now, it is really helpful to look back over the long history of tsunamis to learn how to live with them. Amazingly we can trace the geological evidence of tsunamis far back in time. The oldest we now know about happened almost three and a half billion years ago when an asteroid fell into a shallow sea.

Not bad.

Just to get that in some sort of perspective, three and a half billion years is a longgggg time ago.

If we assume a generation to be about 30 years on average, then most people can expect to have known their grandparents, a mere 60 years separating the generations. Let's go back 5000 years, about 170 generations, and we can find ourselves, for example, in the European Alps with the Ice Man—Ötzi, 5300 years old—the ancestor of millions of people alive today, from our Cro-Magnon past. This is just about imaginable to us, and it is really helpful that we can physically see Ötzi in a museum—strangely the distance of years doesn't seem so much in this case.

This asteroid that generated our first known tsunami is so much older than that, so old that the use of generations becomes somewhat redundant, but just to finish the analogy, three and a half billion years ago represents around 117 million generations, not that humans were around then! These are almost impossible numbers to deal with, and yet tsunamis have been naturally occurring on our planet probably since well before then. We are late arrivals into this picture. But ever since we finally came on the scene, we have been dogged by such catastrophic events and so if we want to learn more about how these things tick, then we have a lot of evidence to sift through.

This evidence in the case of our oldest tsunami comes not from the tsunami itself but rather from circumstantial evidence of the thing that caused it, the asteroid, and some of the after-effects of it hitting the earth that are recorded in the landscape. This evidence takes the form of a "Microkrystite spherule bearing breccia" that now languishes above the ground in the outback of northwestern Australia.

It is here that we face a problem in the telling of the tsunami story—jargon.

We are trying to lay the first building block of the first strand that leads us to our end goal. While it can be most definitely stated that the study of ancient tsunamis is not the sole domain of the geologist, we will spend a lot of time in the geology world and a phrase like "Microkrystite spherule bearing breccia" seems to suggest that the study of ancient tsunamis is indeed geological turf.

Fair enough, in this instance it is most definitely geology that has allowed us to look at this ancient event, so what does this phrase mean? If you cut through the jargon, you end up with a reasonably prosaic interpretation. Microkrystite spherules are small glassy balls of material, a product of the extreme heat generated when a large asteroid hits the earth (another way of saying this is when an almost irresistible force meets an almost immovable object)—basically, the earth melts. The spherules formed by this heat are encased in a rock, a breccia, which is composed of all the stuff ripped up and thrown around by this asteroid impact. It's a bit like a rock soup that is frozen in time.

Asteroids have a great track record of causing some pretty amazing tsunamis (hopefully that is not considered too enthusiastic a word to use in the case of these natural killers). This particular asteroid fell, 3.5 billion years ago, into a shallow sea and so it must have generated a tsunami simply because it relocated a lot of water very quickly. But we don't know how big the tsunami was . . . yet. However, we do know that the 14 km (8.5 mile) diameter Chicxulub asteroid that struck the Caribbean about 65 million years ago generated waves up to 4.5 km (2.8 miles) high, at least according to recent models. That is pretty impressive. The environmental disaster caused by this asteroid is far better known, though, for effectively wiping out the dinosaurs—some of them perishing in the tsunami. And this is where our single strand, the geological evidence for an asteroid that must have caused a tsunami, is joined by another strand, the palaeontological evidence for the chaos caused to the biological environment. Which strand comes first, the chicken or the egg, or in this example, the dinosaur or the spherule perhaps?

We can identify the effects of ancient tsunamis using both geology and paleontology without necessarily finding evidence for the tsunamis themselves. These strands based on circumstantial evidence may seem a little tangential though. What about more direct evidence for the tsunami itself?

I make no apologies for what follows because I am a geologist by training although some colleagues might say I am a geomorphologist or a physical geographer, a geo-archaeologist, a micro-paleontologist, an anthropologist, or possibly even just someone who likes joining up the scientific dots. Whatever, my PhD is in geology, and we have to start somewhere, even though all of the things that I might be have a bearing on this story.

So, the default is always to return to the geology, to the sediments that tsunamis both leave behind and take away. The big key to this work lies in the modern world. "The present is the key to the past" is an often-used adage, but as many geologists like to point out, it is also important to recognize that the past is also the key to the future. It works both ways. We can forensically examine all of the geological characteristics of recent events and use these as a guide to identify similar events thousands to billions of years in the past. In doing so we find events that were significantly larger than anything we have experienced in historical times, tsunamis that undoubtedly lie ahead of us in the future.

The Chicxulub asteroid impact is a good, albeit somewhat extreme, example.

We will return to what an ancient tsunami deposit looks like later on because as all of us are only too aware it is not the geology that concerns us when a tsunami strikes, it is the disruption to the environment we live in, the destruction of our world, and the devastating loss of loved ones.

A large tsunami can uproot entire forests, destroy crops either by ripping them out of the ground or by submerging them in saltwater. Animals drown, buildings are obliterated, entire communities are wiped off the map, our way of life is often changed forever, assuming we survive in the first place. A tsunami might even be called a preferential killer, taking out the elderly and those less able to escape its devastating power, mothers trying to rescue their children, and the children themselves unable to outpace the devastating speed and destruction of the waves. But anyone who lives close to the coast or has their lives intrinsically tied to the sea by employment or lifestyle are at risk (and let us not forget lakes and rivers too).

These are the real reasons why we want to learn more about ancient tsunamis. How big have they been in the past, how often have they occurred, and how can we be better prepared for them in the future?

The questions are both simple and fundamental, and the solutions to them will continue to challenge us well into the future.

Let us not get too far ahead of ourselves. There are a few things that we need to know before we return to the strands. The term "ancient tsunami" is a little vague and scientists have therefore come up with a word—paleotsunami (being British I tend to drop in an extra 'a' and use palaeotsunami, but I am flexible), defined as a tsunami that occurred prior to the historical record. I must admit that I will not stick rigorously to this term in this book but it useful to know it.

This concept of a paleotsunami sounds pretty reasonable, but as we will come to see, nothing here is quite as simple as is seems. For example, one of the most well-known paleotsunamis ever studied is the AD 1700 Cascadia event generated by a massive underwater earthquake just off the Pacific Northwest coast of the United States. The earthquake that generated the tsunami also caused much of the coastline to subside, killing coastal forests, and then the sea inundated the land with a massive tsunami covering everything in sediment-laden saltwater.

Careful dating of trees killed in the event but still standing today as ghost forests along the Pacific Northwest coastline was done using a technique called dendrochronology—a cool tool and based on a really simple premise.

Dendrochronology (*dendron* meaning tree, *khronos* meaning time, and *logia* meaning the study of) or tree-ring dating is based on the fact that most trees grow annually and leave that record behind in what we call a tree ring. New growth in trees occurs in a layer of cells near the bark and in a predictable pattern throughout the year in response to seasonal climate changes, resulting in visible growth rings. Each ring marks a complete cycle of seasons, or one year, in the tree's life. When the tree dies the rings stop growing, so simply counting the number of rings from a tree that died today for example will tell you when it started growing. Over the years dendrochronologists (an impressive job title) have built up huge tree ring databases for different tree species in different parts of the world, many dating back thousands of years. They have done this by looking at trees that are alive today (you take a small core through to the middle of the tree—it doesn't hurt it) and then finding older ones that died say 200 years ago (I once cored an old tree that was part of someone's log cabin—they weren't in at the time!) but that have overlapping ages of tree rings, and so on. Since tree rings vary in width each year depending upon the climate you can, if you are lucky, use this long record of variations in the width of tree rings to find out the precise year that a tree died. A wonderfully elegant and simple way of dating old bits of wood, although be warned, like all things in science, it is not always as simple as it sounds!

In the case of this Pacific Northwest tsunami, the tree rings clearly showed that it happened early in the year of AD 1700. As dating goes for events in prehistory this is about as good as it gets given that there is no historical record to back it up. The tsunami that struck immediately after the land had subsided inundated most of the coast but, just like the 2004 Indian Ocean and the 2011 Japan tsunamis and many before them, it was an ocean-wide event. In this case the waves traveled across the entire Pacific Ocean eventually causing destruction as far away as Japan, some 7500 km (4600 miles) away.

We know this because in Japan it was an historically documented event, and because of this we can actually work out even more precisely when it happened. We now know that the earthquake happened around 9.00 p.m. Pacific standard time on January 26, AD 1700. This was an amazing piece of scientific detective work by both US and Japanese scholars, right up there in Sherlock Holmes territory.

But here's the issue, there is a disjunct here. In one country there are written records and so it is an historical event, in the other it is prehistoric. Is this cheating? Not really. These hybrid tsunamis as they are known are moderately rare events, but they actually help scientists "prove" that what they are identifying through geology as a paleotsunami (one lacking historical evidence) in one country is actually what they say it is, because there is an historical record for it elsewhere.

And this is very important.

As is slowly becoming obvious, paleotsunami studies draw on the expertise of many disciplines, not only geological evidence. For example, the AD 1700 paleotsunami (and the earthquake that caused it) has been identified through many environmental changes such as subsided land, buried vegetation, and drowned trees. It was also recorded in the oral traditions of numerous Native American groups along the coastline, and it has been found in archaeological evidence for abandoned coastal settlements, the loss of food resources, and cultural changes. And yes, it has been found in the geology, through the sediment it deposited and what was in that sediment.

Paleotsunamis can often be part of a suite of landscape responses to something like a giant earthquake or an asteroid. In such an event, the physical environment can be pushed far beyond anything it can cope with, past an environmental threshold from which it will never fully recover, though it will always be trying to recover and that leaves indelible markers for us to read. In the AD 1700 event the land subsided, causing massive ecological chaos that can be charted through to the present day. For example, plants that liked dry soils and had been happy for hundreds of years suddenly found themselves either dead or distinctly unhappy, and wetland loving plants were moving in on their turf—literally.

In some cases, as we saw earlier, we do not even need to find the geological evidence for a paleotsunami to know that one happened. There is a suite of proxy data—a toolbox, if you wish—that can be used to identify whether or not a paleotsunami occurred. We will look at these in later chapters because in a way each proxy is a strand in itself and these strands continue to be created even now as more and more innovative techniques are applied to this search for past events. However, there is a familiar refrain here:

Things are never quite as simple as they sound.

There are some detractors out there who question whether, even with the toolbox available to us, we can truly identify a paleotsunami. How is the evidence, whatever it is, different from an ancient storm, or even an ancient waterspout (yes, they exist too)? An obvious retort could be—well what about the AD 1700 event? We got that right! In hindsight, that was a fairly easy one to identify—not at the time, that was ground-breaking work, but now that the work has been done, we can smack our heads and go "duh" of course it was a tsunami.

Many other deposits are not quite as clear as the AD 1700 tsunami and that is where the toolbox comes in. And as the toolbox grows many of the issues highlighted by detractors are being resolved to the point where it has even become possible to detect "invisible deposits" where there is no sediment to see but only the microscopic or geochemical evidence for the saltwater that covered the land. As such, we are now on the brink of being able to better understand just how big paleotsunamis were, and as I said earlier, that is important.

So, we have a melting pot of different scientific perspectives all contributing to this virtual toolbox. This not only leads to rapid innovations but also to . . . serendipity. And this is a tale of one of those serendipitous events.

Aligning the strands

The strands started to move toward alignment on Saturday, September 24, 2016, at 5:05:54 p.m. Australian eastern standard time with an invitation from an archaeologist in Chile to a geologist in Australia. The hand of collegiality was extended across the Pacific from one country to another and from one science to another.

But why?

Ah well, now there is an interesting question again. How many times do you receive emails from people you don't know and then open them? Of course, there are the obvious junk ones that are immediately assigned to the junk/spam/never want to hear from these people again folder, but there are always one or two that you are not quite sure about. The dithering moment follows, hand poised over the "open" command. Will this assign the entire computer to oblivion or maybe, just maybe, will everything be OK? As my finger hovered in this position of indecision there was just the slightest inkling that I had heard of this person who had sent the email. At this point I didn't know he was an archaeologist, but I did know that the email came from Chile—".cl" did at least give me a hint.

OK, so I had worked in Chile before, and it had only been a year since I had been there doing some poking around in the odd Chilean wetland. But then I had also been receiving an unreasonable number of Chilean spam emails since that visit, courtesy of a hotel that seemed to pass on my email address to all and sundry. Hmmmm, why did the name ring the vaguest bell? One of the benefits of NOT opening the email but actually having the email address to look at meant that I could do a bit of sleuthing before committing one way or another to the dithering finger. The internet can be a wonderfully useful tool and it can also be the bane of one's life, it doesn't have a particularly user-friendly filter at the best of times. BUT at least it does come up with everything it can find and attempts to put it in some sort of order that may or may not be useful.

Bingo! The first result solved the mystery.

Ah ha—he is an archaeologist.
Ah ha—he has done some cool stuff along Chile's coastline.
Ah ha—he works with a Chilean geologist I know.
Ah ha—this is legit.

Feeling slightly idiotic for having thought of this as having been junk mail, I opened it. In time it was to lead into an Aladdin's cave of science, with much bafflement, excitement, and head scratching and many eureka moments, but right now I was just interested to see what they had to say.

To cut to the chase, there was a wonderful statement in the email that said something along the lines of "we are currently working on the hypothesis of a mega earthquake and possible tsunami occurring in northern Chile around 4000 years ago." Could I help?

Now then, now then. A mega earthquake no less, something to rival the largest earthquake in history. This historical quake, in AD 1960, had also been spawned off the coast of Chile. It had been recorded as a Magnitude 9.5 earthquake in southern Chile, but these guys were working in northern Chile where nothing more than about an 8.5 had ever been recorded . . . historically.

It is perhaps worth mentioning here that a Magnitude 8.5 earthquake is no small event. In general, an earthquake with a magnitude between 8.0 and 8.9 will cause major damage to buildings, destroying most of them. It can even cause heavy damage to earthquake-resistant buildings! BUT an earthquake of 9.0 or larger will cause close to total destruction with massive damage to or collapse of all buildings. Shaking, and heavy damage, will extend over a vast area with permanent changes taking place to the ground surface. A 9.5 is not good at all. Magnitude is a number representing the total energy released in an earthquake. The energy released is determined by how much rock moves and how far it moves. For each whole number increase in magnitude, the seismic energy released increases by about 32 times. That means a Magnitude 8 earthquake produces 32 times more energy—or is 32 times stronger—than a Magnitude 7, or in this case a Magnitude 9.5 is 32 times stronger than a Magnitude 8.5 earthquake.

If there had been an earthquake bigger than an 8.5 in northern Chile, then this was a game changer. It had huge significance not only for Chile but for all the other countries around the Pacific that would have been in the line of fire from the tsunami it generated.

I need to take a quick step back here and make a comment or two about the Pacific Ring of Fire. This refers to the edge of the Pacific Plate that is marked by a line of volcanoes (hence the fire) that ring the Pacific. This term goes way back into the 19th century, but in many ways the volcanoes are the obvious bit. They form what is known in the trade as a volcanic arc, or a chain of volcanoes that form above a subducting plate. The Pacific Plate is one of many tectonic plates that "float" around the Earth's surface. When they come into contact with a neighboring plate, one of them normally loses out to another and is pushed down (subducted) underneath it. Usually, a denser, oceanic plate sinks down below a lighter continental one, and as it plunges into the underlying mantle it melts and forms magma that percolates up to the surface forming a line of volcanoes parallel to wherever this subduction is taking place.

It is the invisible, underwater, subduction zone that causes all the grief. The two plates that meet each other are not entirely happy about it and so this is not a smooth, quiet process. One plate grinds against the other, and every now and again there is a bit of give and the denser one subducts and plunges just a little bit more beneath the other. This movement causes an earthquake. The greater the length of the meeting point between the two plates that moves at any one time coupled with how much it moves vertically determines just how big that earthquake will be. In addition, while subducting underneath a continental plate, the oceanic plate pulls the continental plate down a bit by friction. Not surprisingly, the continental plate doesn't like that too much and when it bounces back to restore the status quo, a tsunami is born as the entire seafloor readjusts itself.

Imagine, if you will, around 1600 km (1000 miles) of this meeting point on the seafloor—what most of us tend to call a fault line—rupturing and the seafloor moving about 15 m (50 ft). This doesn't happen instantly and at this scale the "unzipping" can take up to 10 minutes. That is how long an earthquake can last for and that is what happened in the Magnitude 9.1 subduction earthquake that took place on December 26, 2004. Since these types of events happen just offshore, when that vertical adjustment occurs it rapidly pushes up the sea all the way from the seafloor to the surface, and a tsunami is "propagated." In 2004 that tsunami was the Indian Ocean Tsunami that killed over 200,000 people.

The amount of energy released by the 2004 earthquake as it moved both vertically and horizontally was over 1500 times the energy released by the Hiroshima atomic bomb. The total energy of the tsunami waves generated by the fault movement was more than twice the total explosive energy used during all of World War II (including the two atomic bombs).

These are not by any stretch of the imagination little minor blips in the lives of humans that live adjacent to such subduction zones. But they are just that for the earth.

The December 26, 2004, earthquake was "only" a Magnitude 9.1. The Magnitude 9.5 earthquake on May 11, 1960, in southern Chile was about four times stronger, and we know how devastating that was. The tsunami alone killed at least 139 people and destroyed well over 1500 houses . . . in Japan, over 17,000 km away on the other side of the Pacific.

So back to the email.

Here was an archaeologist (and a geologist too) in Chile suggesting that an event greater than a Magnitude 9.0 had happened in northern Chile. OK, so it happened a long time ago, but until the very moment of opening that email no international scientist as far as I knew was aware of this, no one outside Chile had felt the chill of reality, the frisson of fear, the implications not only for Chile but for all of those countries in the tsunami's line of fire.

Those other countries in the line of fire were not the same as those affected by the AD 1960 tsunami. Just where the brutal energy of a tsunami's waves is focused depends upon many things, but the one immediate thing that shunts a tsunami in any particular direction is the orientation of the fault line that has moved. If the fault line (or subduction zone) goes from north to south, then the wave is pushed out west and east at right angles to the fault, if it goes northwest to southeast the tsunami is directed to the southwest and northeast, and so on. Many other variables come in to play from this initial starting point, but the waves head inexorably in the direction they were pushed to start with.

A classic example of this would be New Zealand and Chile. While the AD 1960 Magnitude 9.5 Chilean earthquake (southern Chile) was much larger than the AD 1868 Magnitude 8.5 Arica one (northern Chile), the orientation of the fault rupture where the earthquakes actually took place meant that the direction of the AD 1868 tsunami's main energy spread across the Pacific Ocean directly toward New Zealand whereas that of the 1960 passed mostly to the north of the country (Figure 1.2). What this means for New Zealand is that the largest historical tsunami it has received from Chile came from the northern source, not the southern.

Ah.

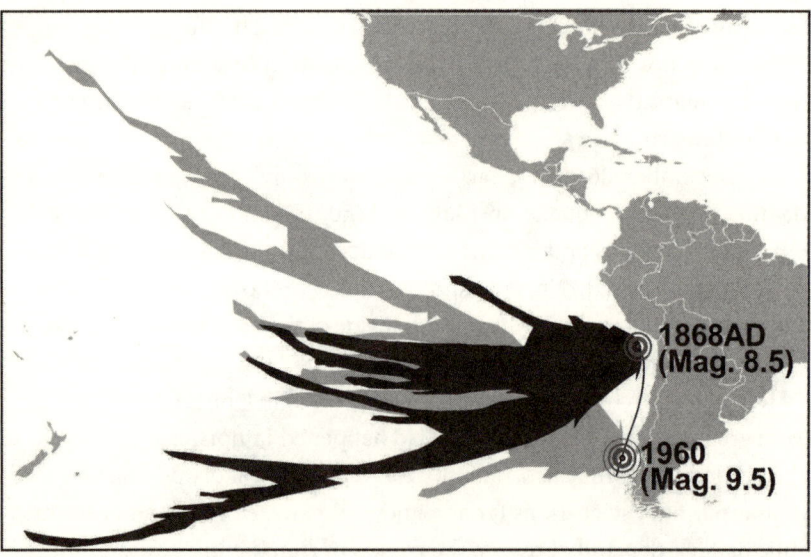

Figure 1.2 The main tsunami energy from the AD 1868 and AD 1960 Chilean tsunamis is governed primarily by the orientation of the subduction zone (thin black line running parallel to the coast of Chile). (Image: J. Goff)

So the worrying point here is that if the hypothesis being put forward by the Chileans turned out to be correct, then 4000 years ago or thereabouts New Zealand saw something much, much bigger than occurred in AD 1868—could it be true? It was also a little concerning to see that northern Chile is home to what some of us consider to be a fairly mature "seismic gap"—a segment of an active fault known to produce significant earthquakes but that has not actually produced one in an unusually long time.

The Chileans (and New Zealanders to be honest) really needed to know whether or not there could have been an earthquake far bigger than had been experienced in historic time, because if there had been then it could be building up to one right now.

This was an extraordinary problem requiring an extraordinary solution. The tsunami toolbox was on hand, but the environment of northern Chile is a huge challenge for researchers. Northern Chile is dominated by the Atacama Desert, a strip of land over 1600 km (1000 miles) long squeezed between the coast and the Andes. It is the driest non-polar desert in the world, with an area of around 105,000 km² (40,000+ square miles) composed almost entirely of stones and sand.

Numerous tsunami researchers from around the world have wandered down the coast hoping to find their paleotsunami El Dorado, applying well-developed geological techniques and coming up blank. Immediately to the north, in Peru, researchers studying how well recent tsunami deposits were preserved in the almost identical coastal desert landscape found that things were a bit grim. Three moderately large tsunamis inundated the Peruvian coast between 1996 and 2007, and their preservation along this arid coastline varied a lot. In one instance all traces of the tsunami have been removed or reworked by flash floods and stormy ocean waves. In another, the wind had removed all the fine sediment, and at another site storm waves had again eroded the evidence.

All that happened to tsunami deposits in just 10–15 years of desert life. What on earth could be expected to survive for 4000 years?

It is important to remember though, that this book is not all about a search for a 4000-year-old paleotsunami, but rather a look at what it is that we do to try and find out whether something of the sort has happened in the past. What lines of evidence can be used and what new thinking can be brought to bear on the problems of working in such extreme conditions? And then we will get to the mother ship—a 4000-year-old event in Chile!

In a sense this is what the toolbox and the nexus of individual strands is all about. It is all very well pottering around your backyard sounding intelligent, but if you really want to test yourself, take yourself to a new place with the worst possible preservation potential in the world, a hostile environment where half of this beautifully developed toolbox of proxy data can simply be discarded because it will never work in a desert, meet a whole new team of colleagues, many of whom are archaeologists, and whose language you don't speak.

A good test.

For want of a starting point, the first strand in this tale is firmly anchored in the present. This is useful because before heading into the past it is helpful to know what you are dealing with and as I have said, the present is the key to the past.

2

Strand 1

December 26, 2004—Indian Ocean

TSUNAMI BULLETIN NUMBER 001
PACIFIC TSUNAMI WARNING CENTER/NOAA/NWS
ISSUED AT 0114Z 26 DEC 2004
THIS BULLETIN IS FOR ALL AREAS OF THE PACIFIC
BASIN EXCEPT
ALASKA—BRITISH COLUMBIA—
WASHINGTON—OREGON—CALIFORNIA.

.... TSUNAMI INFORMATION BULLETIN ....

THIS MESSAGE IS FOR INFORMATION ONLY. THERE IS
NO TSUNAMI WARNING
OR WATCH IN EFFECT.
AN EARTHQUAKE HAS OCCURRED WITH THESE
PRELIMINARY PARAMETERS
ORIGIN TIME—0059Z 26 DEC 2004
COORDINATES—3.4 NORTH 95.7 EAST
LOCATION—OFF W COAST OF NORTHERN SUMATERA
MAGNITUDE—8.0
EVALUATION
THIS EARTHQUAKE IS LOCATED OUTSIDE THE PACIFIC.
NO DESTRUCTIVE
TSUNAMI THREAT EXISTS BASED ON HISTORICAL
EARTHQUAKE AND TSUNAMI DATA.
THIS WILL BE THE ONLY BULLETIN ISSUED FOR THIS
EVENT UNLESS
ADDITIONAL INFORMATION BECOMES AVAILABLE.

In Search of Ancient Tsunamis. James Goff, Oxford University Press. © Oxford University Press 2023.
DOI: 10.1093/oso/9780197675984.003.0002

THE WEST COAST/ALASKA TSUNAMI WARNING CENTER
WILL ISSUE BULLETINS
FOR ALASKA—BRITISH COLUMBIA—
WASHINGTON—OREGON—CALIFORNIA.

International Tsunami Information Centre

It is entirely understandable that the 2004 Indian Ocean tsunami may have slipped into the back of most people's memories or been forgotten completely. After all, the world hasn't had an easy ride over the intervening years. So before we start on the first of our strands it is timely to have a brief recap.

If there is one thing that is worth remembering it is that no one knew that this massive earthquake and tsunami were coming. Even when the earthquake shaking had stopped after 10 minutes, very few people thought that a devastating tsunami was on its way.

There were exceptions though, the most memorable of which was probably an English schoolgirl named Tilly Smith, then just 10 years old and on vacation with her parents at Maikhao Beach in northern Phuket, Thailand. Her school term had recently finished and one of the last things she had learned about in Geography class was what happens when a tsunami approaches the shore. So when the sea began to rapidly withdraw, she recognized that as one of the signs of a tsunami and alerted her parents, who along with resort staff helped to evacuate the beach. Theirs was the only ocean-front resort struck by the tsunami in Thailand where there were no casualties.

Yes, there was at least this one good news story to emerge from the death and destruction of that day.

What came as a surprise to many people was the fact that there was no warning of the tsunami—shouldn't we have known? After all, the Pacific Tsunami Warning System (PTWS) was well established by the late 1960s and had a well-oiled system for identifying earthquakes that might generate tsunamis and quickly sending out warnings to all concerned. Indeed, they issued a tsunami information bulletin a mere 15 minutes after the earthquake— the bulletin is shown at the start of this chapter. This rapid response was based on data collected from a complex network of sea-level gauges and deep-sea sensors linked by satellite to round-the-clock monitoring stations in Hawaii, Alaska, and Japan. BUT, and this is a big but, it was only able to issue a bulletin for all the countries that participated in the network, and not surprisingly it indicated that the earthquake posed no threat of a tsunami to the Pacific region.

The earthquake had occurred in the Indian Ocean.

An hour later, after the size of the earthquake had been revised upward as more data became available and could be analyzed in greater detail, a second alert was sent out by the PTWC warning of a possible tsunami in the Indian Ocean. Frantic phone calls were made to countries that might be affected, but without a list of contacts, and in the absence of an Indian Ocean warning system, it was a lottery. They called embassies, the military, local government officials, anybody. At best the response was disorganized and lethargic. Even the few people who were aware of the dangers were stymied by a lack of preparation, bureaucracy, and inadequate infrastructure. And the tsunami had already struck the coast of Indonesia and was rampaging across the Indian Ocean heading toward Sri Lanka, India, The Maldives, South Africa, and so on.

The scene was therefore set for the biggest disaster of the modern era. The statistics are morbidly fascinating. As briefly outlined in the previous chapter around 1600 km (1000 miles) of fault ruptured, with the seafloor moving about 15 m (50 ft) vertically. This massive movement had knock-on effects, such as causing the entire planet to vibrate by over a centimeter. But the most devastating knock-on effect was the tsunami it generated. This was up to 30 m (100 ft) high and killed an estimated 227,898 people in 14 countries as far west as South Africa on the other side of the Indian Ocean (Plate 2.1).

As science goes, this horrendous event provided an unprecedented opportunity to forensically examine all aspects of a truly massive tsunami. This may sound a little callous, but this one event probably moved our understanding of these hazards forward several decades. Before the 2004 Indian Ocean tsunami a moderately small international group of scientists had been working doggedly and diligently to better understand these killer waves. But in the immediate aftermath of the disaster many science teams, mainly geologists, from countries all around the world mobilized in order to study on-the-ground evidence and to gather a veritable cornucopia of other data.

The importance of gathering these data as soon as possible after an event had long been recognized by the science community. Evidence for even simple things like how the tsunami flowed over the land and eroded or moved around obstacles are invaluable in trying to save lives in future events. For example, what type of vegetation can be planted to slow down the waves? How can buildings be better constructed? This list goes on.

BUT to be able to study the evidence, scientists need to get into the affected area as quickly as possible, at the same time as the humanitarian response efforts are being rolled out. If not, the evidence is destroyed. Diggers move in to clear away debris, vegetation dies quickly and with it evidence of the direction in which it was pushed over by the waves, and so on. In these situations, scientists are in a race against time. The random and ad-hoc collection of data is of no help to anybody. Data needs to be collected in a systematic manner, and it needs to be reported in a way that allows the entire tsunami research community to interrogate the findings.

And so in 1998 UNESCO-IOC (the United Nations Educational, Scientific and Cultural Organization—Intergovernmental Oceanographic Commission) had created an International Tsunami Survey Field Guide outlining the general tsunami and damage survey techniques that should be used in these cases. This manual (revamped in 2014) is useful for all researchers conducting any kind of post-tsunami survey, and so it should be read by all participants. As for the geological information it contains—well, there is a description of how to survey tsunami deposits in the field. Why UNESCO-IOC? Well, they run the IOC Tsunami Programme that aims to reduce the loss of lives and livelihoods caused by tsunamis and not surprisingly, they recognized that geology is an important cog in that machine. The program also gives the science teams that operate under the UNESCO-IOC banner the necessary raison-d'être to be on the spot when the human crisis is unfolding. We are part of the solution not part of the problem. We are not just annoying scientists getting in the way of the humanitarian crisis.

Let us be under no illusion, it is not a pleasant task, and most geologists are not trained for something akin to frontline battle conditions and all that entails. It is a little-known fact that while in most recent catastrophic tsunami scenarios such as this, medical teams and humanitarian aid personnel are routinely offered counseling, geologists are not. Or at least, none on any team I have worked with have been offered such help. Many colleagues have carried out numerous immediate post-tsunami field surveys, and the sites and scenes they experienced leave indelible marks on their minds and can over time build up like barnacles on the hull of a ship. But they still do it because it needs to be done.

Not surprisingly, geological data abounds, gathered by the numerous teams that have entered affected areas in the immediate aftermath of disastrous tsunamis. But the 2004 tsunami was the mother ship, and a massive undertaking. It is difficult to even encompass how much data was collected, and so quickly. As an example, a five-day survey of the western, southern, and eastern coasts of Sri Lanka by a team of eight scientists from the United States, New Zealand, and Sri Lanka gathered data on the elevations of watermarks on buildings, scars on trees, and rafted debris as indicators of the maximum tsunami height. Eyewitnesses were interviewed to determine at what time the waves inundated the coast, the size of them, and the destruction they caused; and sediments laid down by the tsunami were collected for later analysis. This was just one team, working on a part of one country affected by the event. Their rapidly generated results showed that, as expected, the eastern and south-eastern parts of the island were worst affected because they faced directly toward the tsunami-generating earthquake source just off the western coast of Sumatra, Indonesia. Unexpectedly though, the data also showed that parts of the western shores of Sri Lanka, which seemed to be on the sheltered side of the country protected from the waves by the rest of the island, were equally badly affected because the waves "bent" around the island and also slammed into that coast.

One of the more remarkable findings from this short, intensive survey was that human alterations of the coastline had catastrophic consequences for people and property. It may seem obvious, or it may be that hindsight is 20:20, but this example probably goes some way toward showing just how stupid humans can be.

First of all, it is very important to remember that the sea is to be respected. Duke Kahanamoku, famed early 20th century Hawaiian surfer and swimmer, helped popularize the motto, "Never turn your back on the ocean." His reasons were twofold: he wanted people to watch out for the physical dangers of being hit by a wave, and he wanted us to show respect for the ocean. Both of these reasons apply to this example of what not to do to the coastline.

The coast of Sri Lanka is gorgeous, and many a hotel has been built to take advantage of it and provide tourists with an experience they will never forget. It has some iconic locations, such as around Unawatuna where the famous stilt fishermen can still be seen at work, but possibly not for much longer. The 2004 tsunami fundamentally altered much of the coastline so that it can never again be used in the same way. Normally beaches recover even from the brutal scouring effects of massive storms, and Sri Lanka has had its fair share of those during the monsoons. But the effects of this tsunami were almost terminal for the livelihoods of the stilt fishermen.

Faced with a desperate need for food during World War II, caused by a lack of supplies and overcrowded fishing spots, men wandered out just a little farther offshore to get access to more fish. Initially perching on the wrecks of aircraft and ships on the reef, they later set up stilts on which they could sit or stand, and they continue the practice to this day. There were fairly slim pickings for the stilt fishermen at the best of times, but the tsunami changed the entire shape and structure of their beaches so that the fish they used to catch are now further away and out of reach of their stilt perches. However, over the past couple of decades the stilt fishermen have become increasingly popular with tourists, to the point that the fisherman can now get more money by fishing elsewhere and renting out their perches to others who pose as fishermen for the tourists and charge for their photo to be taken. It's a strange world.

Along some parts of Sri Lanka's sandy coast, beautiful dunes often harbor sheltered watering holes frequented by the country's iconic wildlife, such as Asian elephants, leopards, and warthogs. These sites are extremely popular with tourists and so here is the dilemma. How do you create the best tourist attraction, the one that will make tourists visit you and not your competitor? It obviously has to include the wildlife and preferably have easy access to the sea.

Imagine waking up in the morning in your private beach bungalow to watch the magical sunrise from your balcony. Coffee in hand, you watch the waves gently lapping against the shore, mere paces away, tempting you out for an early morning dip. To your left, you look out over the nearby watering hole nestled in the neighboring dunes and are captivated by the sight of elephants, leopards, and warthogs, all sharing a quiet moment of peaceful camaraderie before sunrise and a retreat into the inland forest. Perfect. It couldn't be more idyllic.

And now, step back for a moment and think about this a little more. How is it that you can see the sea when there are sand dunes right next door to you between the watering hole and the sea? The hotel is on a flat sandy plain adjacent to the watering hole but there are no sand dunes in front. Odd. Odd indeed, but that is the price the environment pays for your sensory pleasures. To enable you to enjoy the twin pleasures of wildlife and sea requires some gentle "massaging" of the landscape, nothing major, just flatten out a few of the dunes to open up a gorgeous ocean vista, place all the bungalows on the beach side of the resort alongside a nice restaurant, and put all of the support buildings like kitchens, laundry, and staff quarters at the back adjacent to the access road through the back dunes. And that is it—the perfect hotel for the perfect holiday . . . on nearly every occasion.

BUT, the removal of coastal dunes for hotel construction can have catastrophic consequences. As a tsunami comes onshore, coastal dunes absorb
and even reflect some or most of the energy of the waves. They may lose a
bit of weight in the process as they are pounded by the waves, but even a
small dune can provide a very effective barrier, although maybe only partially. Take some of the dunes away, and the tsunami exploits the gap. There
is little or no resistance to its relentless energy and it penetrates through the
gap, where there is nothing to stop it, expending its energy on anything in its
way—buildings, people, cars, furniture, and so on. In one place in Sri Lanka
only one person survived, a member of the hotel staff far enough inland from
most of the carnage to be able to escape and live to fight another day. Along
the nearby coast the tsunami was rebuffed by the extensive spread of natural
sand dunes and unable to move inland. In one place, the waves were large
enough to push a boat up on to the very top of the dunes, but no further, and
all of the warthogs, leopards, and elephants survived.

Where there were no dunes the waves of the tsunami traveled almost a kilometer inland, finally meeting their match on the slope of an ancient, undisturbed dune. Here, in their last gasp, the waves deposited all that remained
of their energetic activity. The maximum inland extent of the tsunami was
marked by a strandline, a line of all the last remaining debris that it was carrying. As in the regular high tide marks left behind on the beach as the sea
moves in and out each day, here there were fragments of seaweed and broken
shells. But there were also intact shells, recently ripped up from their sleepy
sandy homes just offshore and thrown inland to bake to death in the hot sun.
And yes, there were also the sad, typical trappings of a modern coastline—
plastic bottles, glass bottles, buoys, fishing rope, pieces of wood, and all of
the expected flotsam and jetsam that plague our beaches today. And then
there was the unexpected debris—fish skeletons, human bones, fragments of
outboard motors, shredded fiberglass hulls, sunglasses, shoes, socks, and so
much more, a kilometer inland. Entire hotels had been redistributed over the
coastal landscape, together with sand, mud, vegetation, and everything that
remotely indicated that there had once been signs of life.

This is the first strand, and a starting point. A tsunami deposit is going to be
composed of all sorts of material, natural or otherwise, that comes from the
sea as well as the land that it flows across. When a large tsunami first moves
onto the land it is usually already carrying a lot of sediment, shells, seaweed—
whatever it has picked up from just offshore. Unlike waves generated by massive storms though, a tsunami can also pick up and transport material from

extremely deep in the ocean, such as microfossils from ocean floor sediments up to 1 km (0.6 miles) underwater, because it is the entire water column that is moving not just the wind ruffled surface. In a really big storm, that "wind ruffled surface" can be quite deep, but it is never THAT deep. And then because a tsunami still has a lot of energy to burn as it comes on land, it starts to rip up whatever is in its path—boulders, pebbles, sand, mud, roads, cars, houses, bridges, people, trees, soil—nothing is safe, even if it is tied down.

How long this destruction lasts varies a lot, but invariably each wave very quickly starts to lose energy, and as it does, it drops the heavy things first. If the tsunami had a full range of sediment to transport, then that will usually be redeposited in a logical sequence. Closest to the coast are the heavy boulders, followed by pebbles, coarse sand, finer sand, mud, and finally, the really light stuff, such as leaves, fish, and shells. There are a couple of extra wrinkles to this logical way of doing things. As the waves travel inland, the material deposited not only gets finer, it gets thinner too. Less material carried by the waves and getting finer as it goes inland means it makes up a thinner sheet of material to be deposited. Also, if you were to walk inland over the deposit as soon as it had been laid down you would notice that at any one point it may vary in thickness, filling in a depression or stretched thin over the crest of a ridge, but whether it was a thick bit or a thin bit it generally gets finer vertically as well, as the water gradually slows down.

So, it may not be rocket science, and having told you this it now seems blindingly obvious, but most (not all—there are always wrinkles) tsunami deposits become finer and thinner inland, and they fine-upward (the finer material is on top). This may not sound earth shattering but it is useful in so many ways.

It is important to note that the aftermath of the 2004 Indian Ocean event was not the first time these characteristics were recognized. In fact, a 2001 paper outlined a sort of geological wish list of what scientists would like to find in the sediment record in order to be able to say with some degree of certainty that they had found a tsunami deposit. Even this was based on a lot of earlier work.

BUT the 2004 event was the first time that such a major scientific effort had been put into researching a truly large tsunami in the modern era.

That having been said, the fact that UNESCO-IOC created their sort of cheat sheet for scientists in 1998 was a clear indication that science teams like these had been working on modern events for some time. However, the 2004 Indian Ocean tsunami was a big one. Nothing had really come close to it since the horrendous 1960 Chile tsunami generated by the largest earthquake in historic time, a Magnitude 9.5. And so, in many ways the 2004 tsunami drew the proverbial line in the sand—and scientists all started singing from the same song sheet. This was a watershed for the science. From this point onward, and armed with their hands-on experience of studying the 2004 tsunami deposits, more scientists than ever started searching for evidence of earlier tsunamis all around the world, more scientists than ever gained significant experience in studying these events, and there was a far better understanding of what the geological evidence of tsunamis looked like.

While we were all soon able to point to a textbook example of a beautiful, photogenic, tsunami deposit that bore all the characteristic features that we would like to find—fining inland and upward, we were becoming wiser as to the vagaries of Mother Nature. There are a lot of "wrinkles," tsunami deposits that refuse to conform to a textbook version, and it is all these wrinkles that makes the science of what we do so fascinating.

If you think about it, it is pretty tough to get a nice smooth deposit that shows that the waves gradually slow down as they move across the land, before they gently place their last fine grains of mud and vegetation down in a last gasp before retreating back to the sea again. After all, anyone who has a morbid, or professional, interest in tsunamis will have seen the unprecedented aerial camera footage of the 2011 Japan tsunami rampaging across the ports, towns, and paddy fields of the Japanese coast.

There was nothing gentle, organized, or smooth about it. However, it was a truly textbook example of a large tsunami. Houses, boats, cars, animals, people, trees, everything you could possibly imagine was visible either floating, drowning, or being smashed to smithereens. But, fascinatingly, among this carnage, later geological studies in the coastal paddy fields would find vast areas of tsunami deposit that did indeed fine-upward and inland. Equally, there were vast areas where this was not quite so obvious. Ships perched on top of buildings, vast areas of ground excavated around the sides of resistant buildings as the waters accelerated around the edges, scouring out deep holes. Car tires impaled, spear-like by entire trees, impossibly bent lamp posts, empty beer cans and bottles scattered seaward from breweries destroyed by the backwash (the wave returning to the sea). In Rikuzantakata, in the Iwate Prefecture of northeastern Japan, an entire forest of thousands of trees was uprooted and carried kilometers inland, bar one, sole survivor (Figure 2.1). In other areas boulders the size of trucks were dumped in a field hundreds of meters (yards) inland and left sitting gently countersunk into a

Figure 2.1 Rikuzentakata: The Miracle Pine Tree—only survivor of a coastal forest destroyed by the 2011 Japan tsunami. Note that a high tsunami wall has now been built seaward of the tree, it can be seen under construction in the background. (Photo: J. Goff, 2016)

bed of gravel and sand. At the Port of Sendai, a CCTV camera filmed a telephone box filling up with water as the tsunami rushed inland. When they visited the area later, scientists found that as the water had slowly leaked out of the box it had left behind the perfect tsunami deposit—the coarsest sandy material was on the floor while each successive ledge and nook higher up, even the cradle for the phone, was covered in a dusting of finer and finer sediment all the way up to a little cap of mud perched on the upper lip of door—it fined-upward exactly as it should have done!

Every single one of these oddities can be explained, every single one of these is indicative of what makes tsunamis such remarkable and terrible forces of nature. The most important lessons that many scientists learned in those frenetic few weeks following the 2004 Indian Ocean tsunami became invaluable in understanding what happened in 2011 in Japan, and in turn the 2011 added to the ever-growing compendium of knowledge.

I was once pressed to define tsunami deposits in layperson's terms, or to at the very least attempt to distil in one sentence what they were. It was a difficult question and one that at the time I thought would have benefitted from more than the fleeting seconds offered to the mind in a TV interview. However, in the years that have passed since that interview I have failed to find one that better explains the bafflement sometimes faced by geologists as they attempt to understand what is in front of them—a tsunami deposit is "a deposit out of place."

This is a huge over-simplification, and yet it is true in some way, shape, or form—I think.

BUT, remember, in this book we are dealing primarily with paleotsunamis and not the immediate aftermath of a devastating modern event. But what does this information from the 2004 and 2011 tsunamis tell us?

First of all, we have to step back a second from the geological evidence of a modern event and look into a scientist's mind. This is dangerous at the best of times, but in this instance it is important. If we are looking for evidence of past tsunamis then most, not all, evidence will lie buried in the ground beneath all of the stuff that has arrived and the things that have happened since the tsunami left its deposit behind. As a simple example of this, researchers have returned to many of the sites of the 2004 Indian Ocean tsunami to follow up on their earlier work, to check how the evidence has changed over time. And boy has it changed.

Most, but not all, of the areas affected by the 2004 Indian Ocean tsunami are tropical. Vegetation there grows like crazy, fueled by warm weather and lots of rain. Couple that with the fact that a large tsunami such as the one in 2004 has the unerring ability to significantly upset the physical environment—landforms are obliterated or scattered over the landscape, vegetation is destroyed, and the bare soil and sediment exposed by the waves sits there offering no resistance to wind and water. Until vegetation starts to take root again and the roots help to anchor it in place, the exposed soil can be moved around far more easily than before. In this intervening time, we see the soil establish a precarious equilibrium between itself and the ongoing processes that dominate the area. By this I mean the wind and the rain, and any other weather in the neighborhood. This normally means that the wind blows the soil around until it cannot blow it anymore, or water washes it into a place where it can no longer do much more to it.

This "shuffling" of soil and sediment also applies to the tsunami deposit itself, the newcomer to the environment. After all, the deposit sits on top of the now-dead vegetation that used to grow there and for a time at least, the salty sediments do not exactly encourage the proliferation of verdant plant growth. Nothing grows on top of the tsunami sediments immediately after they have been deposited. It forms a layer of sediment over the landscape, and rain, rivers, wind, and eventually new plant growth will all do their bit to erode it and cover it up. The roots of new plants break up and disturb the tsunami sediments, wind can blow sand around until it starts to form dunes, and water will often carve channels and gullies into the deposit and flushes parts of it back downstream to the sea.

As the years go by these processes can, in certain environments, completely wipe the geological evidence for the tsunami off the face of the planet. Of course, it does rather depend upon where you are in the world. Outside the tropics for example, in places where sediment and soil rapidly build up on top of the tsunami deposit then that deposit can be preserved largely intact. These are the two ends of the spectrum—total destruction or near total preservation.

It will be of no surprise to learn that most cases lie somewhere in between the two extremes. For example, where you have had a deposit of pebbles, sand, shells, and so on, all carried inland by a tsunami, it will be sitting there on the surface of the ground exposed to all of the processes that normally operate in the area. So imagine if you will a coastline with beautiful sand dunes, a lovely sandy beach, not a single sunworshipper in sight, free of even a single tourist. This is not a dream, New Zealand is full of such beaches as are many other parts of the world, but let us take New Zealand and a place called Henderson Bay. It you are ever lucky enough to visit New Zealand then drive north from Auckland, and keep driving. Drive for 160 km (100 miles) and you arrive at Whangārei, you still have a long way to go. Drive another 160 km (100 miles) and you arrive at Kaitaia, the most northerly town of any note. Drive another 60 km (37 miles) and you arrive at Henderson Bay, a mere 65 km (40 miles) south of Cape Reinga's famous lighthouse, a tourist spot that is invariably touted as being the most northerly point of

New Zealand's main islands—it is not, that honor is reserved for North Cape, an impossibly tortuous journey of over 30 km (20 miles) east and north a bit along logging roads and then, well, sort of roads, and walking, and across Māori land to which access must be requested, and granted. So basically, Cape Reinga will do. It is amazing how far north you can travel from Auckland—a trip that is well worth it because the farther north you go the more remote and beautiful the land becomes.

Back to the beauty of Henderson Bay. A tsunami struck here back in the 15th century. We know this because we have done a lot of work on this event, which affected much of New Zealand, but at Henderson Bay we see two wonderful extremes of what the effect of a tsunami can look like. A few hundred meters offshore in about 30 m (100 ft) of water lie the remains of an old pebble beach, a relic of lower sea levels, and this together with a lot of sand, seaweed, and pumice was picked up by the tsunami and carried inland up onto the top of 32 m (100+ ft) high sand dunes, over the top, and into the wetland behind. And there it sat. The wind blew, the sun baked, and the years passed. Some 500 years or more later along came a couple of geologists to poke around and upset the serenity of the site. The tops of the dunes are covered in pebbles, a veneer that stretches for hundreds of meters and is an impressively awesome sight. The pebbles have been blasted smooth by the wind on their upper surfaces but retain their original lumpy rockiness on their protected undersides. Many have been smoothed and shaped into what are known as Zweikanters or Dreikanters, types of ventifact, wind-shaped rocks that normally form in deserts from the abrasive action of blowing sand. Drei, two-edged, and Zwei, three-edged, pebbles are shaped like this because of differences in wind direction. The most obvious smoothed face represents the dominant wind direction, while the less smooth or faceted surfaces represent the next most common wind directions. These pebbles had been sitting on top of the dunes being sand blasted for years while the rest of the tsunami deposit had simply blown away!

If you think that finding pebbles on top of 32 m high dunes is pretty impressive, you only need to travel about 40 km (25 miles) almost due north to Tom Bowling Bay, the bay closest to North Cape, and you can find pebbles from the same event up to 42 m (140 ft) above sea level! And if you really want to scare the hell out of yourself then travel about 450 km almost due south and you arrive at the west coast of the Waikato region. Here at the gorgeous Ngararahe Bay I have traced pebbles up to 65 m, possibly 70 m (230 ft), high on the sand dunes there—a different tsunami, but an impressive beast nevertheless.

The pebbles at Henderson Bay can be traced down the back slope of the dunes but then disappear as they near a wetland formed inland behind the sand—the tsunami's energy was running out of steam, after all it had taken a lot to get over those dunes. A short sediment core taken vertically into the wetland revealed the nice, equally photogenic parts of the same tsunami. At the bottom of the tsunami deposit was a contorted and deformed layer where the wet peaty material of the wetland had been suddenly overloaded by a load of sand and other things being dumped on top of it. Immediately above that were small bits of gravel, sand, and pieces of soil that had been ripped up from the ground that had once covered some of the sand dunes. Above this was the finer, less dense material. Front and center was a lump of pumice, porous, bubble rich volcanic glass, a wonderful abrasive for cleaning hard skin off your feet! This was the sea-rafted Loisels Pumice that has washed up on many New Zealand beaches in the area. It dates to around 600 years ago and so must have been there before the tsunami struck. Above this was the organic material, bits of grass, soil, and shell fragments, all that was left behind as the tsunami's energy waned, and it finally came to a standstill. All in all, a really wonderful example of how just one tsunami deposit can vary across a very short distance after it has been sitting in the landscape for a long time (Figure 2.2).

 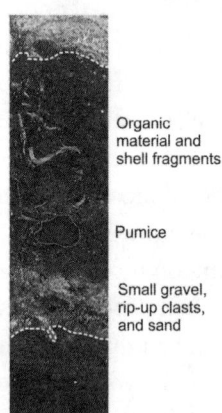

Organic
material and
shell fragments

Pumice

Small gravel,
rip-up clasts,
and sand

Figure 2.2 Henderson Bay, northern New Zealand: Left image—the pebble veneer draped over the sand dunes up to 32 m high. Right image—the deposit in the wetland, fining-upward and giving an idea of when the tsunami happened, which must have been after the Loisels Pumice arrived at the beach. (Photos: Scott Nichol; with permission)

The trick is to know, or at the very least to have a gut feeling about what might be left behind by a tsunami and where it might have left it—this is the "art" of the science.

The art is a bit of a no-brainer. You go for the lowest hanging fruit—the place where you think you are most likely to find evidence for a past tsunami. In other words the best thing to do is dig a large hole (a trench) along the edge of a wetland close to the sea where there is a nice peat or a nice fine organic-rich mud. It is better though, if the peat or mud is not so waterlogged that you end up creating a large swimming pool. In the absence of a decent wetland any low-lying, "quiet" environment just back from the coast will do. You can even wander along the edge of a large meandering river that is quietly making its way to the sea over the last few hundred yards or more of its journey. Pebbles on dunes are fun to find, but they are unusual. I wouldn't go out deliberately looking for them, but they are very cool when you do find them.

The premise here is that a tsunami is very high energy and will carry much coarser sediment (even boulders if you are lucky) into these quiet sleepy backwaters, and that coarser sediments will stay there because there is not enough energy in the normal run of things to move it around at all.

Sometimes though it is simply too wet to dig a nice big hole—a Sisyphian task where every spadeful of material moved from the hole is rapidly replaced by water gushing in. However, there is a seemingly endless variety of high or low tech coring devices that can extract sediment from deep underground. As a very rough rule of thumb about 1 m (3 ft) of soil/sediment (going down vertically) is equivalent to about 1000 years of sediment deposition. I can now hear a massive outcry of disagreement from my colleagues, but I will reiterate—this is a very general rule of thumb, after all a tsunami deposit 10–20 cm (4–8 inches) thick is laid down almost instantly, that's up to 20% of your 1000 years deposited within a matter of minutes. So, this rule of thumb varies a lot but in general it means that if you have a tool that can get you say 6 m (20 ft) of vertical sediment from underground then that is hopefully enough to give you a good idea of what has been going on at that spot over the past few thousand years.

It also means that you have lots of fun getting wet and muddy in the pursuit of science! In many cases though, it is absolutely impossible to be choosy about where to dig or what to use, and you have to do what you can with what you've got to get what you want.

A perfect example of this was when we were working in Samoa a few years after the country was devastated by the 2009 South Pacific tsunami. The aim of the work was to find out answers to the common question—how often had the country been affected by tsunamis in the past and how big had they been?

But I feel that I am getting ahead of myself here as we enter the practical fieldwork side of the science. So let us step back for a second. There has already been a reasonable amount of talk about fieldwork, measuring things and taking samples all in a very organized and "scientific" manner. However, there has been little mention of the person doing the work, the geologist.

What is a geologist?

Most people immediately think of a slightly portly gentleman dressed in khaki, rock hammer in hand, hitting rocks. Well true, there are quite a few of them, but fortunately geology has long stepped away from being the sole preserve of the male, and also it has grown beyond just rocks.

At this point I feel we need an aside within an aside because it is necessary to mention an important point that relates to what geology is today, but something many practicing geologists would disagree with. I have quite simply stated that geology has grown beyond just rocks. This is true, BUT in many academic institutions and other scientific bodies there is the great divide—no, not the big drainage divide in the United States where rainfall on the western side of the divide flows to the Pacific Ocean, and rainfall on the eastern side flows to the Gulf of Mexico—but the great divide between Hard Rock and Soft Rock geologists. This may seem petty, but trust me, it is not.

This basic divide in the discipline can be best explained as follows. Hard rock geologists study igneous and metamorphic (crystalline) rocks like granite and gneiss, they are hard, you hit them with a hammer to break bits off. Soft rock geologists study sedimentary rocks, such as limestone and shale, which are easier to break apart, and generally this is the preserve of those wandering into exploration for oil, natural gas, or coal, although it can also involve those who study sediments *before* they have become rock! In simpler terms, hard rockers use hammers, soft rockers don't.

In reality, while this divide is alive and well, the second group, soft rockers, have expanded into a wonderfully diverse set of subdisciplines and so if asked, may not even be able or want to identify a hard rock. Tsunami geologists pretty much all fit into the second group, not always though, when you get back a few million years even the tsunami deposits turn into rock that may need a whack with a hammer from time to time.

OK, back to the original aside. Naturally, any geologist, or more specifically tsunami geologist, has their own "kit" that they use for their fieldwork.
This comprises their personal field equipment and clothing. There is another
entirely separate set of equipment that would be used by any group in the
field, such as drill rigs or boats, and so on. No, what sets each geologist apart,
a physical representation of their idiosyncrasies, is their own personal outfit.
Invariably it is the well-established "oldie" that has the more weather-beaten,
field-season-hardened gear, often viewed by younger researchers with a mixture of incredulity tinged with sneaking admiration. While an individual's
ensemble may often look ramshackle, dated, and eccentric, everything in
that kit comes to hand immediately when it is needed. No need for endless
searching around for a pen, or a compass, or whatever, they are where they
are and they are always there.

Imagine being Indiana Jones, whip at your side, fedora on your head, and
notebook at the ready.

The loss of a key piece of field gear after many years is like losing an old
friend. I was once working in Indonesia and had by far my most favored set
of field gear I have ever had. I have never been able to replace bits of it like
for like. A faded orange shirt, long sleeved, hard wearing, two chest pockets,
that when new used to shine embarrassingly like a beacon in the dark, but
had mellowed into simply the best shirt for all field occasions. I would not
admit it, but it was on its last legs, or arms as it happened. The stitching was
going, and it was dangerously thin in places. It had done 10 hard years in the
Asia-Pacific region with climates varying from the brutal tropics to the sub-
Antarctic freezer.

I was checking out of a hotel en route to a field site and was, as usual, carrying my field backpack with assorted essentials—water, trowel, trenching
tool (collapsible spade), waterproof, sample bags, notebook, and my Japanese
Nejiri Gama (a small hoe used by gardeners to cut through the top bit of soil
to scrape off shallow rooted weeds and mosses—used by geologists to scrape
off sediment on the vertical side of a trench to create a nice clean surface to
study). The bag was, as usual, fairly heavy.

Could I please sign the checkout form before leaving? No problem,
I swung the backpack up onto my right shoulder and it slide straight down
my right arm pulling the sleeve with it to the end of an era. My shirt was
irreplaceable—no longer manufactured. It has taken me many unsatisfactory
years to find a reasonable replacement, but it is just not the same.

Field trousers—again, I had the perfect ones, khaki, lightweight cotton, not the absolutely useless artificial fabrics with unzipping legs—ugh, useless, if for no other reason than the zip is in the wrong place. You kneel down to examine something and land right on the zip. My field trousers had pockets for Africa and plenty of space to allow you to kneel down or sit without ripping them. I thought ahead and bought two pairs of them once I had got through the first ones. They took a pounding and lasted maybe a year or so each. And when I finally needed replacements—gone, no longer manufactured! Again, the short-sightedness of the manufacturers had failed me. A total tragedy—in the space of about a year the fundamental "me" had gone, lost to discontinued lines of clothing.

It has been a long road, but slowly and surely I have established the new "me." Gone are the lightweight khaki trousers, replaced by heavier duty grey ones. The pockets are still there and the comfort. They can take a pounding and last much longer than the old ones. The shirt, hmmm, well that is grey too, with the pockets, but it is an artificial fabric of some kind and not quite up to the old one, but I tend to wear a battered T-Shirt underneath it and keep it unbuttoned so that's fine. There are really only three other pieces of field fashion that count—the boots, hard wearing, battered, which basically have been around the block a lot. There is absolutely no point in using a new pair because of the pounding they take and so I wear them into the ground. They are not really broken in until the grip on the soles is almost non-existent and the laces broken and retied in a couple of places. The ultimate sign that they are truly field boots is when they are held together with duct tape—that utterly essential piece of field equipment that always travels with me. It serves many purposes—patching holes in trousers, holding boots together, sealing up sample bags if needed, holding my fieldbook together—more on that on a minute—and sealing up cardboard boxes of samples so that nothing, absolutely nothing, can get into them without the use of a decent knife.

The two other pieces of essential clothing—a hat and a bandanna. The hat is an Akubra, battered, sweat stained, holey, and it fits. If it is hot, you can dunk it in water and slap it back on your head, if it is windy it doesn't budge, if it rains, you are dry. The bandanna though is possibly my favorite piece of clothing. It is either tied around my neck to stop sweat dripping down my back or it is on my head when I am digging a hole—the Akubra is all well and good, but not much help in a hole when the brim catches the side of the trench. I have always used Canteen bandannas and have a nice collection of them, although my favorite is a black one. Canteen is a charity that supports young people with cancer, and boy does it do a good job. I have been following them since the early 1990s and always buy a new bandanna if I can on National Bandanna Day. A worthy cause, and always a reminder of how lucky I am when I am out there in the field.

OK—the field notebook. Again, it is the personification of a geologist. What it looks like, what make it is, how you fill it in—pencil or pen—and what you put in it. Mine? Yellow, hardback, Rite in the Rain—yes, the pages are pretty waterproof and a pencil always works whereas a pen is a dicey option. Many geologists are brilliant artists as well, their drawings of outcrops and landscapes, fossils and fine details of rock or sediments are a marvel. Sadly, I have never risen to those dizzy heights, but my childlike scrawls and idiosyncratic shorthand have always worked for me, if for no other reason than no one else can make sense of them so my pearls of wisdom and ponderings remain locked in my indecipherable scrawl. There is more to a notebook though than its make and what you put in it, it is also what you put on it. For example, where do you keep your pencil (or pen) and your spares because you are always guaranteed to lose one sometime? My secret weapon is duct tape—that ever present, completely and utterly essential piece of field equipment—have I mentioned that already? A convoluted wrapping of duct tape around the notebook creates a very convenient pocket in which to slide your pencils so there are always there, always ready, not lost in a pocket or perched behind your ear waiting to fall off. But, just like the metamorphic rock some of my colleagues' study, the patterning and structure of the duct tape wrap has morphed over time reflecting the pressure (intensive fieldwork) and heat (tropical study areas can be demanding) of the job—the basic premise is that there can never be too much duct tape holding everything together! And finally, the piece de resistance—elastic bands, two of them. One is a page marker that allows you to open to book immediately to the current page in use, the other holds the entire book closed. All very convenient, all very "organized," all very "me." No one else has one like it.

A couple more items are worth an honorable mention. Although technically not field kit in the sense that they don't see battle on the field of play, they are essential, at least for me. First cab off the rank is the coffee plunger mug. One of the greatest things ever invented for when you are staying in digs miles away from anywhere, whether with a big or small group of people, when the accommodation is invariably not quite the Ritz but rather a little more down to earth. Whether you are lucky enough to have a room to yourself or are sharing a bunk bed with an inveterate snorer, the one thing that gets me going in the morning is a strong black coffee—or two.

For some inexplicable reason many of my colleagues think that field-work means suffering for the cause, but I draw the line at coffee. If all else fails, and it has on occasion, I will be forced to drink some black swill made with powder or granules—instant junk that first gave me the taste for coffee when I was young, but from which I rapidly matured to beans and a serious need for the blend I like. Yes, a decent supply of ground coffee usually finds its way into my travel bag. Overseas travel can make things problematic though—for some reason the customs officials don't seem to like my Ziplock bag full of ground coffee. Can't understand why. On such occasions, desperate times call for desperate measures and so a game of Russian Roulette is played in attempting to source a local supply, often with disastrous results and the inevitable default of powdered hell. But where ground coffee is available then the mug is a godsend. Coffee plungers are few and far between in most field accommodation and so a mug with a plunger fills not only that role but also at the end of the day when aching bodies are relaxing and the wine starts to flow, the absence of glasses—another accommodation nightmare—proves no obstacle as the plunger is removed from the mug and the drinking begins.

And finally, the Swiss army knife. Yes, on occasion it gets used as a scale in a photo so it could technically be considered a true piece of field equipment, but to be honest, it has a corkscrew and a bottle opener. Need I say more? The privilege of being a scale in field photos is usually reserved for my old faithful United States Geological Survey scale given to me many moons ago and heck, a bit of advertising for my colleagues doesn't go amiss. However, scales can take many forms and be a source of much amusement later when giving conference presentations. In the case of Samoa, from which this has been a long aside, we used beer bottles—Vailima beer to be precise. It was not that we were drinking on the job, but rather that the supply was seemingly endless, washed up in the sad detritus of the recent tsunami—empty.

And so, back in Samoa, our team of "oldies" ventured into the field, com-fortable with the task ahead of us. There wasn't a lot of wiggle room at the selected site—the only possible site as it turned out was between the coast and the cliffs and a trench had to be dug in the available space. Scattered palm trees peered down at the chaotic scene below. The area had been hit hard by the 2009 tsunami, there was wood, endless wood scattered across the scene, broken tables and chairs, smashed up pieces of cabinets of indeterminate origin, walling, buckled bicycles, and rusty nails everywhere—the detritus of a once idyllic life ripped apart and exposed to the world.

This was not a pleasant task, but it was important. A mechanical digger was squeezed into the site and completed in five minutes what usually took us a day of hard digging—this was a luxury. True to the twists and turns of digging holes for tsunamis, it quickly became apparent that teasing the information out of the earth was going to be no picnic. A little over half a meter (1.5 ft) down we hit the water table. In the absence of a water pump to remove the water a simple but effective solution was adopted. A massive hole was dug at one end of trench to act as the deep end to this putative swimming pool. This allowed the water to pond there while the rest of sides of the trench could be studied for tsunami sediments. True to form, and rather pleasingly at the same time, two blindingly obvious paleotsunami deposit layers came into view full of all of things you might expect, that had been picked up from the sea and deposited on the land. This was not only good news, but also served to massage the egos of those of us who had selected the site.

The trench was quite long and because of that the deposits could be seen to be getting thinner and finer inland, but the swimming pool was filling in fast. Speed was of the essence and so the lower of the two layers was sampled and recorded as fast as possible to the constant drone of the mechanical digger as it bailed out the water from the deep end. It was a slow, increasingly wet, losing battle. And just to add to all of this activity, these efforts were all being captured on film for a possible documentary series. It never made it to air, but that didn't make it any less intimidating to have the camera watching everything that was going on that day.

Inexorably, the water level started to rise, ever faster as the tide came in.

Scrape, scrape, scrape went the trowels and *Nejiri Gama*, click went the cameras, and scribble went the pencils as the notebooks inevitably got wetter and wetter. Pencils were lost to the muddy swimming pool, hastily grabbed sandwiches were nurtured carefully in washerwomen's hands, mud and sand slowly seeped into every orifice and collected happily on field clothing both wet and dry, and all was captured for posterity by the documentary team.

As a minor aside—a few days ago while writing this manuscript I received an email from the leader of that documentary team. While they had failed to sell the idea to any TV company the rushes were still preserved on DVDs— would I like to have them? Over 10 years after that day, I had the dubious pleasure of seeing a younger me gradually being submerged in the lukewarm water of the Pacific while standing in a trench doing what I love—being a mud-ologist!

The upper tsunami layer was a mere inch or two above the rising water by the end of the day. The mechanical digger, exhausted after its day-long battle with the sea sat Canute-like in submission as the water swirled around its tracks. Things were getting critical. The trench wall was sagging, and three geologists were still standing in it immersed up to their armpits, and then the rains came. A mixed blessing—yes, it started to wash off some of the mud, yes it was warm, but the trench walls weakened further and the note taking and sample collection became a nightmare of swearing and fumbling fingers. Duct tape and indelible pens came to the rescue as the only things that would hold the written words of the sample numbers on the bags, and then finally it was done.

Leaning back against one side of the trench we admired our handiwork and the completion of a monumental task in record time against all the odds. Just at that moment a foolhardy visitor wandered in to view and enthusiastically jumped in to get a better view. Ill-prepared in flip flops, shorts, and a Samoan shirt they sank beneath the water, and surfaced coughing and spluttering. The small tsunami generated by their actions saw the trench walls finally give in, collapsing at the deep end first and moving with frightening speed toward our relatively safe haven at the shallow end. In a matter of seconds, we were all out, heaving our bedraggled visitor out of the slurry of sediment and water that was threatening to suck him under. All were safe, and with the added delight of not needing to fill in the big hole that we had dug.

Everything was wet, everything was caked in mud and sand, everyone had a strange smell about them that couldn't quite be identified but that was undoubtedly strongly linked to the chaos of the recent tsunami debris that was piled all around us. Yes, we had had all of the jabs, no we were not concerned, although the ever-present rusty nails and threat of tetanus still hung over us.

These were fun days in the Pacific Islands before health and safety ran amok, although to be honest it is not the islands that are the problem, it is the administrators back at base camp who seem to try every possible way of stopping you doing fieldwork. Back in 2009, once we were in the field there was no shoring up of the sides of the trench, no gentle steps down the side for ease of access, just good old fun getting dirty and doing the work that needed to be done, and the denouement was yet to come.

The day finished with a ceremonial walk to the beach a mere 50 m (160+ ft) away. Mud, sand, and smell followed our weary bodies as we wandered trance-like across the beach and into the azure blue sea streaming a trail of muddy water behind us. There was no slackening in the pace until, waist deep, we sat down to reward our bodies in our massive saltwater bathtub. There is nothing quite like sitting in room temperature water savoring a few Vailima beers, watching the slowly setting sun and a growing mud slick around us as the sea gradually eased the mud from saturated field gear.

The end of a tough day at the office. It might not have been a great place to dig, but it was worth it in the end.

And now back to the strand and the geological evidence.

Armed with the knowledge that tsunami deposits are a bit unusual, it starts to become all too easy to "find" them. In the perfect scenario you find a lovely squishy dark brown peat a good meter or more thick, which has a series of lovely clean, coarse sandy layers cut across it like some weird type of multilayered sponge cake. They are not only very photogenic, they preserve the catastrophic effect the tsunami had on the land. This is also "tsunami by color," as the sediment changes suddenly from dark brown peat to bright yellow. Having said that I have seen brilliant white tsunami deposits in New Zealand and red ones in Sri Lanka, so bright yellow is by no means all a tsunami has to offer. The join between these layers is often so marked that in some places, if you are lucky, you can see the remains of crushed vegetation overwhelmed by the waves. Now that sounds nice and easy, who needs to be a scientist to find them?

Well, quite naturally, that is one end of the spectrum. The other end of the spectrum is far from easy. What happens if your tsunami deposit is a nice yellow sand comprised of sediment that has been picked up from the beach and it sits on top of . . . a nice yellow sand that used to be a beach? Sand on sand. Which is tsunami and which is not? This is where geology and all its little sub-branches of expertise come into play.

For example, within every deposit there are microfossils—microscopic life that was living in the sea or in the sediments at the time the tsunami picked them up. These get picked up along with the sediment and carried inland. Of course, in any normal storm or a good wind, a lot of these things get carried inland as well, so the trick is to know which ones were put there in an instant by a tsunami and which have had a more arduous journey to their final resting place. But just to add more complexity, when the tsunami comes onshore it will also pick up a lot of those microfossils that had a more arduous journey to get on land in the first place and will mix them with all the new ones—ugh.

Long arduous hours sitting over a microscope and trawling through lots and lots (and lots) of samples can slowly tease out the differences between the sandy deposits—a tsunami emerges from the murk. It isn't glamorous but it works . . . most of the time.

Most of the time. To a certain extent that is the point though. While we are forever in search of tsunami history, it is inevitable that sometimes you simply cannot tell if it was a tsunami that put the sediment in the ground. This we have come to accept. BUT we can say that if it wasn't a tsunami then it sure as hell must have been one mother of a storm, and while in most cases these are nowhere near as bad as a tsunami, they are still worth knowing about.

Surprisingly, it is often the simple things like tracing how far inland the deposit goes and tracking changes in the thickness and size of the material (those little indicators that are so useful) that will usually throw a storm out of the window and come down in favor of a tsunami. But at the tough end of the spectrum, a tsunami deposit and a storm deposit can be so similar that even the most experienced tsunami researcher will have moments of doubt and uncertainty. It is these equivocal deposits that create a lot of uncertainty and help propagate arguments about whether you can tell the difference between a storm and a tsunami in the geological record. Well, to be honest, in 99% of the cases you really can tell the difference, it is just that the hard ones take a lot more effort to figure out. Many scientists tend to shy away from such efforts, and that is fine. They err on the side of caution, or some simply go for the low hanging fruit sites—the sand in the peat. Fine again too, it is all good work that needs to be done.

Samoa was one of those cases where it was pretty easy to identify the tsunami deposits, but then the field conditions would have put off all but the most hardened fanatics.

Let's take a look at another example of suffering for the science—the Dog's Breakfast Deposit. This can be found in New Zealand, and at only about 500 years old is prehistoric there. The name becomes apparent when you know the context. Take a sleepy coastal lagoon, Ōkārito Lagoon, almost entirely full of grey mud, not quite the sticky kind though, because it has enough slightly coarser silty material in it to make it almost bearable to walk on. Some preliminary cores drilled 2 or 3 m (6–9 ft) down through the mud showed that there was a weird layer of coarse sediment about 50 cm (1.5 ft) down.

Hmm—what could that be?

There was no river nearby that would have been big enough to have put it there, and anyway there was a lot of bits of shell in it ("shell hash" to those in the know). True, the wetland was sitting right on the edge of a stormy coastline, but on the other hand the sea immediately offshore was very shallow for a long way out to sea so not even really big storm waves could get too far inland—they would break a long way offshore. By the time the waves reached the coast, they could easily cause some damage but would not have enough energy to do much more.

Could this odd layer be evidence of an ancient tsunami?

The only way to find out the nature of this beast was to dig a bit of a trench. Fortunately the coastal lagoon was very shallow. You could wade through the water up to chest height at high tide while at low tide it was a gentle, slightly sticky, stroll across the mudflats.

More context is needed here though. The local residents are wonderful, with an eager thirst for more knowledge about their environment. The lagoon is adjacent to an old gold mining town, now a small village with 20–30 people or so. People still fossick for gold here, with each new storm throwing up a veneer of denser black sand and the odd bit of gold onto the upper beach.

The wildlife is a wonder. Surrounded by native forests, the area is home to a unique species of Kiwi (the bird, although some might say the people too) and many other rare native birds. The waters host the famous New Zealand whitebait and other fish.

New Zealand whitebait are the juveniles of the *galaxiid* species which live as adults in freshwater rivers and streams. Most of these species are endangered, and so there is only a short fishing season to catch them during their migration from their larval stage at sea to maturity in freshwater environments near the coast. The most common whitebait species in New Zealand is the Inanga, caught in the lower reaches of rivers or lagoon entrances. Fishers mainly use large, open-mouthed, hand-held scoop nets to catch their prey during the annual migration. Why fish for them? Well, they are a delicacy, and until you have tried a whitebait fritter do not even think to challenge that statement. Most are eaten locally, but for those with an eye to making money it is the most expensive fish on the market, when available. Prices can vary between NZ$70 and NZ$130 per kg!

As if this paradise could not be more special, the lagoon sits within the Westland Tai Poutini National Park that covers a vast area of the beautiful West Coast of New Zealand. Although, a word of warning, while stunningly beautiful this vast area is often called the "Wet Coast" and is immortalized in a now famous poem with an anonymous, but humorous, author:

> It rained and rained and rained
> The average fall was well maintained
> And when the tracks were simply bogs
> It started raining cats and dogs
> After a drought of half an hour
> We had a most refreshing shower
> And then the most curious thing of all
> A gentle rain began to fall
> Next day was also fairly dry
> Save for a deluge from the sky
> Which wetted the party to the skin
> And after that the rain set in.

Rain is not the only eccentricity of this paradise. Once visited you will never forget the sandfly, *te namu* (little devils) in Māori, vampires of the genus *Austrosimulium*. There are assorted Māori *pūrākau* or oral histories about these beasts, but the general gist with regards to the West Coast goes something like this. The god *Tu-te-raki-whanoa* had just finished creating the landscape of Fiordland, but its beauty was so stunning that it stopped people from working and they stood around staring in awe. The goddess *Hine-nui-te-po* became so angry at these unproductive people that she created the sandfly to bite them and get them moving again.

Sandflies roughly resemble gnats, with only two of the dozen species found in New Zealand being the biting kind—the New Zealand blackfly (*A. australense*) and the West Coast blackfly (*A. ungulatum*), and even then it's only the females that bite. BUT, swarms of them descend upon the un-suspecting tourist (and that includes scientists) with locals, mysteriously seeming to be almost immune. When European settlers first arrived on the West Coast they had to smother themselves with rancid bacon fat as a deter-rent . . . joy.

Travel writers place the sandfly even higher up the wickedness ladder than Alaska's mosquitoes. I have experienced the sandfly on many occasions. It is possible to be encased in clouds of sandflies within just seconds of stepping out of a car or in my case, re-emerging from a West Coast lake after going on a sediment sampling expedition on a lake bed. With my wetsuit on I was not bothered at all, but the moment I started to remove it, they descended. Sandfly bites hurt, and they generally manage to latch on undetected. As if that is not bad enough, they release an anticoagulant that keeps the blood coming while causing itching at the same time. Fingernails get filled with blood, and the itch is almost always worse on the second day. These ankle biters from hell just add another layer of misery to a tough day in the field. Most annoyingly, they don't really come out in great numbers when it is raining, but then it is raining and that is miserable enough, but as soon as the rain stops for however short a time, back they come. Relentless, unstoppable, insatiable. How do you combat them? DO NOT wear dark clothes— blues, browns, and black are terrible colors to wear. Eat garlic, drink lots of beer, and smother yourself in the strongest DEET (diethyltoluamide) you can find. Having said that, DEET was developed for US military use, and so you can imagine how good it is for you. In Canada, where they have the black fly, a slightly less annoying ankle biter, they used to sell DEET in bottles up to 99% pure—I loved it. However, in the early 2000s they reduced the maximum concentration to less than 30%. The only side effect I ever noticed though was its ability to act as a solvent—not to be worn with many synthetic fabrics—that actually melted the leather on the inside of my Akubra. What it was doing to me I will never know. The best local West Coast alternative to DEET is Ōkārito Sandfly Repellent. While many people use citronella oil, with varied success, this one uses citronella oil and sweet almond oil—and it works. Perhaps that is why the locals seem to be almost immune.

And so, armed with a multitude of ineffective insect repellents (later expeditions were more savvy) and field gear, two brave geologists wandered out into the unknown. The main problem was that the mud flats are huge and only exposed for about two hours either side of low tide giving a four-hour window for work. Coupled with the fact that it was too soft to get a mechanical digger out there, a form of scientific yomping was required through a couple of miles of knee-deep water, laden with all the digging tools that could be carried. Yes, the digging would be done by hand.

At this point it is important to remember that this meant digging a trench a few meters long through gripping mud to try and get some sense of how the deposit varied inland and also going down well over half a meter (1.5 ft) deep to get beneath the deposit. This was no mean feat. Every spadeful of mud fought to stay where it was and once on the spade fought equally as hard to not leave its new home. And it became more belligerent as the trench got deeper.

The four-hour window was carved in stone, after that we would have constructed yet another seawater swimming pool, albeit smaller, in another country!

Two meters (6 ft) long, 40 cm (1.3 ft) deep—two hours gone. Backs aching, arms like lead, spades feeling like lead weights, half man half mud half sandfly bites, the mud pile grew like an Iron Age ring ditch hiding the world from view . . . and the tide had started to turn.

It is amazing how clothes, however hardened to fieldwork, become increasingly useless as their mud content increases. Weighed down with mud, trousers start succumbing to gravity with desperate grabbing at belts and hitching up merely helping to redistribute the mud inside the trousers, and with the mud came the occasional sand fly, and with the sand fly the occasional bite, and with the bite the endless scratching, and with the scratching the mud moved from the ground, to the spade to the hands to the clothes to the body in an endlessly repeating cycle of hell.

By the time three hours had passed the only way to distinguish between the mud and the person was that the latter was moving, a constant plunge of the spade into the ground, followed by a professional tennis quality grunt as the offending mud was heaved out of the trench closely followed by a thud as the spade hit the mud pile to be scraped clean to return to its task.

At 60 cm (2 ft) deep, bemudded and complete, the task was done. A brief halt in the proceedings was marked by a council worker-esque leaning on of spades and a chance to straighten the back and see the world again.

It is perhaps at this point that a couple of minor, although potentially major, issues came to light. We were in the middle of one of the country's most pristine, glorious, and protected wetlands. The scenery was stunning, and across the foreground of the view at that moment appeared a flat-bottomed jetboat moving inexorably toward the trench—just the kind of boat that could cope with the shallow waters of the lagoon. Two things immediately became apparent—they were deliberately heading toward us and they were in a boat. Possibly a better way of saying this was, who the hell was it and why was the water level so high?

To answer the second part first, the water was so high because the tide was still coming back in and the large mudbank we had built up was the only thing between us and the water. The trench had not filled in yet because the water took a little time to filter through the mud. Speed was of the essence! And that is where the second point comes in. There were a few small issues: (a) we had forgotten that it was World Wetlands Day; (b) where else better for a little piece-to-camera by the Minister of Conservation for national TV than a nice world famous pristine New Zealand wetland? (c) we needed a permit to dig (we eventually got belated permission), and (d) the entourage was heading our way because the local guy driving the boat wanted to see what we were up to.

I won't say we were freaking out at that moment, but there was a gentle panic beneath the surface veneer of the fake Hollywood smiles. It will always remain a mystery to me, but the minister was effusive and the cameras were not even rolling at the time. What were we doing, what had we found, how was the digging, where were we from? Great to see the science being done . . . blah, blah, blah—and then it was over, they were off, and two stunned mullets stood there trying to work out exactly what had happened and why we were not on our way to prison. The only explanation we could come up with was that his underlings either didn't want to tell him we were digging an illegal hole, or they simply assumed that we had a permit. Whatever the correct answer was, speed was still of the essence. We didn't want to be around when they came back.

As if to let us know our time was running out, the bottom of the trench was already permanently wet, and the level was going to start rising fast very soon. But the deposit was exposed in all its glory. A mass of shell hash was the first sign of a tsunami, infilling gaps between a thick layer of intact marine shells. These were filter feeding shell species that had literally been ripped out of the seafloor, separated from their homes, and thrown into the lagoon. Being lighter than the sediment they lived in they had been rafted on top of all of the coarser material and deposited in the last gasps of the tsunami as the energy waned enough for them to settle out along with the shell hash, bits of seaweed, and the odd bit of wood.

Austrovenus stutchburyi, the New Zealand cockle, *Tuaki* in Māori, an edible bivalve that generally lives in the subtidal or lower parts of the intertidal zone about a couple of centimeters (a few inches) down in the mud. Just offshore were the perfect conditions for them but they had been seriously rehoused. Amazingly after hundreds of years they were still intact, their two opposing shells were still joined, their long-desiccated abductor muscle faintly visible—apparently it had imparted such a strong control over the shells that even in death they refused to be parted—the only difference seemed to be that one second they had been alive, the next they were part of the tsunami. Prizing them open after 500 years—yes, they still needed to be prized open—we found the desiccated pulp of the mollusk still there, protected, unblemished, cocooned in its sedimentary grave. Amazing.

Beneath the shells was sand, beach sand, but beneath that was gravel, and deeper down, albeit all of this change happened over only a few centimeters (inches), were large pebbles picked up by the tsunami from 30 m (100 ft) or more below the sea level and a long way out to sea (Figure 2.3).

Figure 2.3 Dog's Breakfast Deposit, West Coast, New Zealand: more a photo of mud, a muddy foot, and a hole that looks like a grave for a guitar. At the top end of the guitar is a sloppy mess of shells and gravel collapsing into the trench before a decent photo could be taken. (Photo: J. Goff, 2004)

This is really where the difference between storms and tsunamis becomes noticeable, or at least the difference between a big tsunami and a big storm. The depth to which a storm wave can move sediment or even affect the seafloor in any way is called its wave base. This is equal to half of the wavelength—the distance between waves—so the longer the wave, the deeper the wave base. During severe storms, really severe storms, wavelengths may even reach as much as 250 m (800+ ft), but that only gives a wave base of 125 m (400+ ft), well short of the potential of tsunami waves. As for the largest storm waves ever experienced, there is evidence from Trinidad Head lighthouse in California that on December 31, 1914, it was struck by a wave that could have been as high as 61 m (200 ft). The only eyewitness, Captain Fred Harrington, was the keeper of the lighthouse and he noted:

> The storm commenced on December 28, 1914, blowing a gale that night. The gale continued for a whole week and was accompanied by a very heavy sea from the southwest. On the 30th and 31st, the sea increased and at 3 p.m. on the 31st seemed to have reached its height, when it washed a number of times over (93-foot-high) Pilot Rock, a half mile south of the head. At 4:40 p.m., I was in the tower and had just set the lens in operation and turned to wipe the lantern room windows when I observed a sea of unusual height, then about 200 yards distant, approaching. I watched it as it came in. When it struck the bluff, the jar was very heavy, and the sea shot up to the face of the bluff and over it, until the solid sea seemed to me to be on a level with where I stood in the lantern. Then it commenced to recede and the spray went 25 feet or higher. The sea itself fell over onto the top of the bluff and struck the tower on about a level with the balcony, making a terrible jar. The whole point between the tower and the bluff was buried in water. The lens immediately stopped revolving and the tower was shivering from the impact for several seconds.

The lighthouse stands on a steep cliff 53 m (175 ft) above the sea. Since the water fell over the top of it, the wave may have been 61 m (200 ft) high. Impressive, but sadly not verified in any record books. In the well-known Guinness World Book of Records, the largest recorded wave was 25 m (84 feet) high and struck the Draupner oil platform in the North Sea in 1995. As an aside, the largest wave ever ridden by a surfer was a 24 m (80 ft) wave off Nazaré, Portugal.

Of course, these all pale into insignificance in comparison with the largest historically documented tsunami—the 524 m (1720 ft) high Lituya Bay wave of 1958! And if you really wanted to take the boat out (perhaps an unfortunate turn of phrase in this context), then mathematical models of tsunami waves generated by asteroid impacts into the deep ocean—as and when this happens—can essentially be as deep as the ocean is at the point of impact, up to around 5 km (over 3 miles) high or more!!!

So picking up pebbles from 30 m (100 ft) below sea level doesn't seem that tough now, although to be able to pick them up AND carry them inland over (or through) a bunch of sand hills, does need the type of sustained energy provided by the extraordinary wavelength of a tsunami than can be many kilometers long. However, even then, being heavy the pebbles are some of the first things picked up by any tsunami to get dropped or deposited—and here we get back to the fining-upward of the deposit. As the energy of the tsunami wave decreases, smaller and smaller particles of sediment are deposited, and the finer sediment also gets transported farther inland too, so yes—this deposit conformed to that model we have come to look for—it fined upward and inland, and yes, everything in it came from the sea with the exception of a small mix of dune sand picked up when the tsunami blasted through the dunes on its way into the lagoon. In among the fine organic rich layer mixed in with the light and fluffy shell hash on the very top of this deposit there was undoubtedly the remains of some of the vegetation that had been growing on the dunes, and even some of the soil ripped like the shells from its resting place.

From the deposit we found in the trench wall we could see the story of destruction unfold. This was no storm.

And then, without warning, the trench collapsed.

In an instant the perfect structure and organization of the deposit descended into chaos and the whole thing looked like . . . a Dog's Breakfast.

With little else to do other than let the water reclaim what belonged to it, it was simply a case of packing up the equipment. The trench was a now a long, water-filled grave-like structure, a shadow of its former self, but still a clear blot on the remarkably pristine landscape. A few more tides would see it returned to normal though. Sodden, muddy, and facing the arduous task of man-hauling a ridiculously large number of samples across the kilometer or more of mudflats that were rapidly becoming a lagoon again, we turned to face the waters . . . and the boat reappeared on the horizon.

Was this a miracle or had the troops returned to exact their vengeance?

The sole occupant was the beaming captain who was returning to the scene of our crime to see how things were going. Ostensibly, he was offering us a lift back to the shore, which quite frankly was a godsend, but we suspected that beneath the affability lurked an ulterior motive. A true local, the captain had a thirst for information and knowledge about the lagoon—he could now talk more wisely to his clients about what lay both beneath and around the lagoon.

This was not an opportunity to miss. In exchange for a ride home, a few hours of quiet conversation in front of a roaring fire, and a growing collection of empty bottles of Marlborough Sauvignon Blanc, he milked us for information. This was an exchange that was willingly accepted—after all we were visitors and would most likely never return, and he lived there, working the land and the sea in a remarkably environmentally sensitive manner. Making a living helping tourists enjoy one of nature's great gifts to our senses is a remarkably enjoyable pastime, and to be a small part in helping that experience by adding the odd pearl of geological wisdom was a pleasure. We were all happy with our day's work.

But to drag us back to the thread of this chapter, the point is that yet again, the geology worked.

This coastal lagoon site was one of the first of many sites along the West Coast of New Zealand that point to a catastrophic paleotsunami affecting the region some 500 years or so ago. We didn't really think of it as the first site at the time. We were thinking that we needed to make a really convincing case for it actually being a tsunami deposit, that was key. After all, as the 1979 saying coined by Carl Sagan goes, "Extraordinary claims require extraordinary evidence." This needed to be a good site with good data to support our claim. And it was. Our day on that lagoon laid the foundations for the detailed description of an event that is now supported by evidence from over 40 different sites. To put this in context, when the site was first studied there was little if any enthusiasm for the suggestion that New Zealand had any evidence at all of prehistoric tsunamis, especially since historic ones all seemed to have been a bit of a damp squib. The findings from the Dog's Breakfast were therefore a serious wake up call, but to be honest it was a bit of a no-brainer. New Zealand sits astride the Pacific Ring of Fire, not even next door to it but right on top of it, and to even think that there was no tsunami hazard bigger than had been experienced in its incredibly short historical record (less than 200 years) was a tad naïve. Having said that, it has taken a long time to steer the supertanker of conventional science thinking in the direction of a more realistic recognition of the country's position in the world. Twenty-five years later the debate about just how bad the tsunami hazard is for the country goes on—and that is good because there is now a debate. At the very least there is now a cadre of early, middle, and late career experts capable of recognizing a tsunami deposit when they see it—a quantum leap over a mere 25 years. This is wildly encouraging.

While the Dog's Breakfast tsunami was a prehistoric event it did actually affect humans, humans who used oral records and not the written word to tell of such events. Oral records are a fascinating component of the tsunami toolbox, but we will not venture too deeply into that side of things in this book. Studies of oral records are books in themselves. Each oral recording needs to be treated with both caution and respect because the true meanings of these traditions rest with the indigenous people that they belong to. As outsiders it is both unwise and disrespectful to color such stories with our own interpretations, especially if they are massaged to fit the story we want to tell.

Many a scientist has wandered this path to their peril. Let's pay a short visit though because it is important.

3

Strand 2

Same Time, Different Places

Mātauranga Māori [Māori knowledge] is a valuable and neglected area of information and understanding about past catastrophic events . . . we map Māori oral traditions (pūrākau) that relate experience with extreme environmental disturbance (in particular, tsunamis) . . . compare the findings with geoarchaeological evidence, and discuss the scientific benefits to be gained by considering pūrākau as legitimate perspectives on history.
D. N. King and J. R. Goff. Benefitting from differences in knowledge, practice and belief: Māori oral traditions and natural hazards science.

Oral recordings, oral traditions, traditional environmental knowledge, indigenous knowledge—they have been discussed under many guises with each new term somehow appearing to try and take the moral high ground by using some form of "politically correct" terminology. Ultimately these are messages, often multilayered, with threads woven together to sustain a richness of information that is largely unavailable to the outsider. Engagement with the people who genealogically link to such stories requires close attention to a politics of representation, in both past recordings and current ways of retelling them.

In simple terms, important messages were passed on orally and this oral record changed with the teller and listener. Different threads lead in different directions but within the whole there is a core that holds the threads together. In the case of tsunamis, it can often be the event itself that is the core—a catastrophic event of such social and cultural significance that it binds together the other threads.

In Search of Ancient Tsunamis. James Goff, Oxford University Press. © Oxford University Press 2023.
DOI: 10.1093/oso/9780197675984.003.0003

I do not even suggest that what I talk about here goes any way toward describing the complexity of each oral recording, but what I do feel is that the kernel of some records is represented by a tsunami, a tsunami so devastating that its effects remained immortalized over 500 years after it happened in the memories of cultures separated by thousands of miles of the South Pacific ocean.

Western science likes to think that it is best. We have the best equipment— the best machines that go "ping," the best analytical techniques, the best minds, the best education systems, etc., etc., etc.

The hubris is monumental.

Over 500 years since a devastating paleotsunami rampaged through the South Pacific a few Western scientists have dug a few holes in a few places, and following a lot of head scratching and a lot of analysis carried out using a lot of different techniques, they have finally—with the exception of a few researchers who disagree largely because they are not tsunami experts— decided that there is geological evidence for what must have been a very large tsunami.

Big whoop.

In parallel to this major geological research, a few archaeologists had also started scratching their heads about things that happened around 500 years ago to try to explain some strange similarities in the dating of settlement pattern and cultural changes recorded in the archaeological record across vast swathes of the Pacific. Prior to this attempt to join up the dots of region-wide archaeological evidence a variety of different hypotheses had been put forward to explain each individual case. Some of the explanations that gained traction were definitely plausible at the time, but times move on. For example, widespread abandonment of coastal sites, resource depletion, and warfare were thought to be the result of the Little Ice Age (LIA), as cooling temperatures made it impossible to grow certain crops. In the Southern Hemisphere the LIA generally happened between about AD 1450 and AD 1850, continuing well after the arrival of Europeans and the start of written records.

However, there is no obvious indication of such crises at the time of European arrival, with perhaps the exception of Easter Island, but that is a more complex story. But the timing of the start of the LIA fits with around about the time that rapid changes appear in the archaeological record. The idea that the LIA was responsible was proposed from about the 1970s onward and fit well with exciting new work coming out of the Northern Hemisphere on the effects of the LIA. However, this explanation does not really fully address the abruptness of the changes that took place and why people moved away from the coast. The instantaneous abandonment of coastal sites across significant lengths of a country's coastline and in different countries in the South Pacific is not to my mind explained by the LIA hypothesis.

Still, the narrative of the effects of the LIA has continued to live, thrive, and survive well into the 21st century. Of course, it was developed in the 1970s and it was not until the 1990s that the geological evidence for prehistoric tsunamis really started to be understood. This gave a good 20 years or more for the LIA hypothesis (among others) to get imbedded into the scientific psyche. So in some ways it is not surprising that it has proven so difficult to dislodge it even as the science has moved on—there are still those who are yet to be convinced about the new kid on the block, the tsunami hypothesis, and that is fine. It is just like a group school report for the tsunami community as a whole saying, "We need to work harder."

Such minor disagreements between disciplines are commonplace though and no big deal. More significantly, western science in general has proven remarkably adept at doing its own thing in an increasingly myopic fashion, comforted by the knowledge that with new toys (equipment) comes an enhanced ability to find out even more "stuff" from really small amounts of actual on-the-ground physical evidence, and that must be good, right? Sadly, the answer to that question is both yes and no. It is always good to learn more. BUT with an ever-increasing technical sophistication, every individual science discipline has the ability to countersink itself deeper and deeper into its own myopic view of the world, producing more and more data that requires more and more analysis and interpretation, and focusing in on more and more detail at the expense of anything else.

Yes, this is great for advancing the science and getting a far more detailed picture of something, but it is catastrophic for trying to see the bigger picture, the one that takes all the detail of that something and looks to see if it is similar to any other things out there and might even help join up the dots between things across disciplines. The general refrain goes something like, "Sorry, we are far too busy looking at our own little thing in even more detail to worry about what others are doing."

This is not a new observation by any stretch of the imagination, so please do not think that I am simply nitpicking over some local bunfight. Oh no, there are many classic examples out there—let's have a look at the continental drift debate as just one example.

Continental drift is the hypothesis that the Earth's continents have moved or drifted over geologic time relative to each other. First speculated about in AD 1596 by Abraham Ortelius, it was more fully developed by Alfred Wegener in 1912. Wegener first presented his hypothesis to the German Geological Society on January 6, 1912, suggesting that the continents had once formed a single landmass, called Pangaea, before breaking apart and drifting to their present locations. While he presented a lot of evidence for this drifting of continents, he had no convincing explanation for an actual process that might have caused this to happen, other than possibly by the centrifugal force of the Earth's rotation.

While the guts of this hypothesis are now accepted to the point that it underpins our understanding of the science of plate tectonics, it was rejected for many years. Apart from anything else, Wegener was not trained as a geologist but was a meteorologist and a pioneer of polar research. His hypothesis was met with skepticism from geologists who really considered themselves the experts in this sort of thing and therefore viewed Wegener as an outsider. They were also extremely resistant to change.

Wegener died in 1930 well before his idea was proven. Indeed, in 1939 an international geological conference on the subject of continental drift was held in Frankfurt to really deal with it and cast aspersions on the whole topic. This conference was dominated by the fixists—those who thought that the continents did not or even could not move. While this may seem laughable today, it is interesting to know that when Sir David Attenborough recounted that when he attended university in the second half of the 1940s, he "once asked one of my lecturers why he was not talking to us about continental drift and I was told, sneeringly, that if I could prove there was a force that could move continents, then he might think about it. The idea was moonshine, I was informed." It was still being refuted as late as 1953, but by this time the tide was turning as huge amounts of data were emerging in support of the hypothesis.

The Swiss geologist Caesar Eugen Wegmann refused to teach the ideas of Wegener, maintaining that the kind of sterile and pointless debate which those ideas engendered should be avoided in favor of focused research on solvable problems. While he considered continental drift to be a myth, he also recognized that the question of whether it was a myth could not be solved by geology alone, but rather by multidisciplinary research linking geology and oceanography.

In a sense Wegmann was sitting on the fence, but he had a good point. Multidisciplinary work was what solved the problem and proved Wegener's theory, but it involved a lot more than just geology and oceanography. Wegmann was to a large extent fed up with groups of scientists writing papers that all agreed with each other and simply piled one paper on top of the next adding a little more detail each time but sticking safely within the same paradigm. At the end of the special issue of the Geological Review (*Geologische Rundschau*) of the Geological Association of German and International Scientists that reported on the 1939 Frankfurt conference on continental drift, he inserted a cartoon that is as pertinent today as it was then. It shows the men of science, each peering intently into one of many narrow tubes or cylinders clinging desperately to his (yes, just the male perspective in those days!), "point of view" (*die wissenschaftlichen gesichtspunkte*). Wegmann's message, while clearly highlighting the myopic views of geologists, also shows that real satisfaction comes when you step back to look at the whole and not just a single piece.

And so, in the context of our journey here—there was geological evidence for a region-wide paleotsunami in the South Pacific some 500 years ago or so, and this experienced push back from some in the science community. OK, then let us as a tsunami community "work harder." For starters we can look to the discipline of anthropology . . . enter oral recordings.

Of course, oral recordings had been known about for hundreds of years by those in the know—the indigenous peoples whose history these represented—but at best they were only of a vague peripheral interest to those studying tsunamis, after all they were just stories, not real science. Although it is fair to mention that it was not as if geologists and archaeologists were working in a void here. Victorian ethnographers around much of the colonial world had been avidly collecting what were variously called Native Stories or Native Myths. The collection, translation, and transcription of these oral records reached a highly competitive level in some countries with individual "researchers" even paying indigenous people to tell them more stories that they could then publish and become more famous than their colleagues—a quick turnover being favored over accuracy of translation. Needless to say, indigenous groups were as savvy as anyone else and so many "new" stories were created in order to fill the order book! While this showed a shrewd appreciation for the value of a good story, it also led to later scholars being remarkably dismissive of entire collections.

Unfortunately for tsunami research the convergence of Victorian ethnographers in search of a good story coupled with missionary zeal to convert any "native" to Christianity, led to the creation of many Great Flood stories "à la Noah."

Despite this unfortunate coincidence, many fascinating tsunami-related oral records turned out to reflect what happened remarkably well, as revealed when subsequent geological work was undertaken. For example, Lieutenant George Thornton Emmons of the US Navy spent quite a lot of time in Alaska in the 1880s and 1890s maintaining law and order. During his time there he bonded well with the indigenous Tlingit Indians gaining much respect from them and for them. In the process he journeyed with them in their canoes, bought numerous artefacts, and transcribed many of their oral recordings.

One of those he recorded tells of the monster of Lituya Bay who dwells in the ocean caverns near the bay's entrance. He is known as Kah Lituya (the Man of Lituya) and he resents any approach to his domain. All of those whom he destroys become his slaves, and take the form of bears, and from their watch towers on the lofty mountains of the Mt. Fairweather range above the bay they herald the approach of canoes. With their master they grasp the surface of the water and shake it as if it were a sheet, causing tidal waves to rise and engulf the unwary.

This, along with many similar stories, would no doubt have been consigned to spending its life as a humorous anecdote told at dinner parties in various parts of the colonies if it wasn't for one thing, or rather THREE things. Tsunamis in Lituya Bay in AD 1899, 1936, and 1958 would take this story from mythdom into reality many times over, and as it turns out there had been earlier tsunamis too. This oral record underpins the truth behind what is still the largest historical tsunami on record, the 524 m high (1720 ft) behemoth that struck Lituya Bay in 1958.

So, oral records are rubbish? They have no bearing on reality?

Now we can read on with a little alarm bell lodged in the top of our skull and see what we make of the next few interesting examples from the South Pacific. I could spend a long time explaining that they all relate to an actual tsunami, but hopefully you will accept it when I tell you that they do even if I only offer a short explanation.

Before we discuss oral recordings from the South Pacific—Vanuatu, Wallis and Futuna, Cook Islands, and New Zealand—we can definitely say that we now know from the geological record that a tsunami related to these long-standing traditions did actually occur. Among other things, there was a huge volcanic eruption that generated a tsunami around AD 1452–1453 in Vanuatu and a massive earthquake just to the east of Tonga around the same time. For the archipelago of Wallis and Futuna we know from geological AND archaeological records that there was a devastating tsunami around the same time, AND from New Zealand we have a plethora of geological and archaeological data that tell us the same thing. These countries are thousands of kilometers apart and yet we have evidence for ONE big event that was catastrophic to the people that lived in those places. The exception is Vanuatu and its volcano where the oral recording refers to a local tsunami around the same time—but hey, geologically it looks like Vanuatu could have been hit by the bigger Tongan one as well as the local tsunami, so who knows, several Ni-Vanuatu (indigenous Melanesians) stories may have been merged or they might have been unlucky enough to have been struck by two tsunamis within a matter of a few years (Figure 3.1).

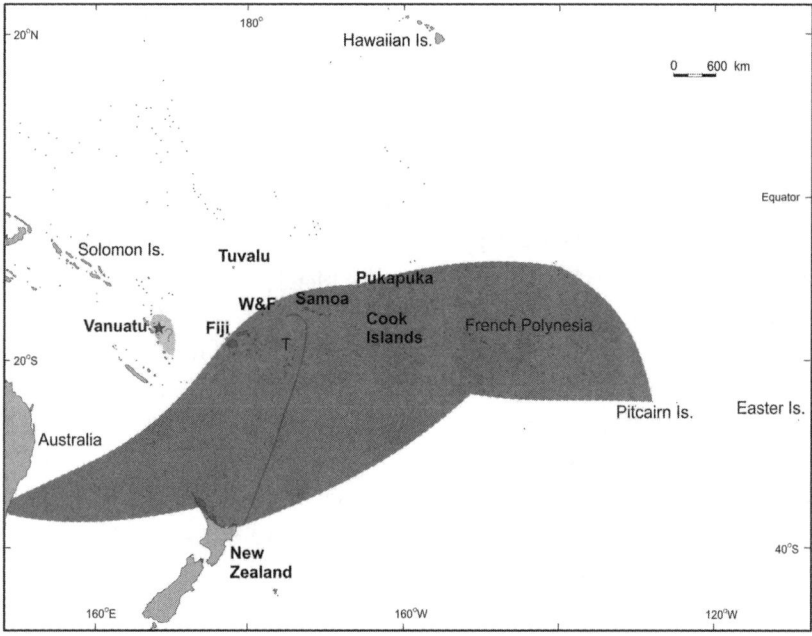

Figure 3.1 Southwest Pacific—Current estimated extent of two mid-15th-century paleotsunamis. Light grey shading around Vanuatu: Kuwae eruption sourced event; asterisk: Kuwae volcano; Dark grey shading: Tonga trench earthquake sourced event; grey line extending from New Zealand to Samoa: Tonga trench; T, Tonga; W&F, Wallis and Futuna. (Image: J. Goff)

And while geologists and archaeologists have taken decades of the modern era to unearth enough data to get to the point where the respective scientists from each discipline can now at least discuss how bad or relevant these tsunamis are to the human story, indigenous cultures from around the entire region have been aware of paleotsunamis for hundreds of years!!!

Isn't Western science marvelous?!

So let's delve into indigenous knowledge. It seems to be worth dipping our toes carefully into those waters.

The Legend of Kuwae volcano is an oral recording from the island of Tongoa in the Vanuatu archipelago. The story has been taped and stored in the Vanuatu Kaljoral Senta (Vanuatu Cultural Centre). It was told in Bislama by Chief Tom Tipoamata of Tongoa. Bislama, also known by the French name, *bichelamar*, is a creole language, and one of the official languages of Vanuatu. More than 95% of Bislama words are of English origin, with a few French and other sundry words from various languages of Vanuatu. It is a wonderfully adaptable language that moves with the times, so for example a "bra" is "basket blong titi" in Bislama.

The legend of Kuwae volcano is as follows (I have abbreviated it a little):

Kuwae [the volcano] is said to have originated from the island of Tongoa. Long ago, in the centre of Vanuatu, between the islands of Epi and Makura, there rose up a big island and here and there, there were little volcanoes, extinct volcanoes and peaceful volcanoes. This island was called Kuwae. Years and years ago, on Kuwae, there lived a great man, a tall and very strong man. He had long hair and a bushy beard and a splendid necklace of shells around his neck. On his arms, pig's tusks and brass bracelets accentuated his muscles. This man was called Pae. On the whole island of Kuwae, he was known to be the best bow shooter. He was very talented, and he never missed his target. Pae was a very proud man. While hunting, it was always he who killed the best pig. While fishing, the best fish were for him. No one could come close to Pae. This of course, provoked jealousy. Pae was admired but not liked.

"He is always the one to kill the best pigs [and] the biggest fish are always on his line!"

The people of Kuwae became tired of Pae's prowess. One day, a group of young people got together.

"Friends, do you not find him too proud?"

"Yes, the girls look only at him, everything is for him! I think we should play a trick on him that will make him ashamed to show his face in public."

"What a good idea! Let's go!"

After this discussion, the youth from Kurmambe village decided to act. That evening the nakamal (traditional meeting place) was dark. There was no moon, only a small, dying fire in one corner of the building. The boys went to get the girls and bring them to the nakamal. they were all there laughing and joking. On Kuwae, women were allowed to enter the common meeting place. They could even sleep there.

Laughter echoed in the dark, and a woman slipped onto Pae's mat. Pae felt the female body next to him and he entered her. Suddenly, the mood was broken by a loud cry.

"Ah! My son! My son! I have just slept with my son! Ah!"

The woman had just touched a raised scar on Pae's chest. She then ran away leaving Pae speechless.

"My mother! I slept with my mother! Oh horrible incest!"

Alas, it was too late because in Pae's head, the idea of revenge had already formed.

"I will have revenge. They'll see who they're dealing with!"

One day, the people from Ikatoma and Anabong decided to go to the southeast of Ambrym where they had some family. To make this voyage, they had built an enormous canoe. Pae thought that the best solution was to leave for Ambrym, the country of fire. His uncle lived over there, and his uncle knew everything necessary for him, the perpetrator of incest, to take revenge on the others. So he left with the people of Ikatoma. Once they arrived on the beach, his uncle, who had just seen the canoe arriving went to meet him and upon seeing his drawn face, asked what was wrong with him.

"My uncle, the people of Kurumambe tricked me and I slept with my mother! It was their fault! Their fault!"

"My son, I understand your anger at the trick they pulled on you. What do you want to do about it?"

"I want to take revenge. I want the fire of my fathers . . ."

"The fires! Pae, my son, think twice, think of the innocent living on Kuwae."

"Uncle, I want my revenge."

So Pae's uncle agreed and told Pae that the next day, they would go and get what Pae wanted. And the next day, the men climbed high up into the rocks where nothing grew and a sinister smell of sulphur pervaded the air.

"Here?"

In front of them was a strange hole. It was very hot and there were lizards around it. The uncle then called the spirit of the volcano and said to his nephew, showing him the big lizards swarming around the hole.

"Which one do you want? This big one?"

"No, it is too big I will not be able to carry it."

His uncle then showed him a smaller one that was similar to the common blue-tailed lizards that are often found warming themselves on the beach.

"Yes, that's the one I want!"

"But Pae, this lizard is too powerful. It is capable of destroying everything!"

"That's the one I want!"

"Alright Pae, if that's what you want. However, I am warning you that it could destroy everything at once. Here, take this yam and dig a hole in the inside. Put the lizard in it and close it back up. Close it well."

Marakipule, the man who had rowed the canoe and brought Pae to Ambrym came back to get Pae. Pae attached the yam containing the lizard to the mast of the canoe still known as Warasole. When they were halfway to Kuwae, the lizard's power was so great that it split the canoe in two. Marakipule, the rower, did not understand what had happened. The canoe continued to stay afloat, however. Water could be seen in the crack in the canoe, but it did not come in the boat. Warasole continued its trip. When they had arrived at Kuwae, Pae went and buried the yam containing the lizard at the foot of an ironwood tree, next to the nakamal. Then he covered it with a big stone and said to Marakipule:

"Leave quickly! Take your canoe and go out with your family to Efate. Direct your canoe towards Emao Point where we can see on clear days, called Aken. When you arrive at Aken, you will see another point on Efate called Maniura, which is next to Forari. Go there and stay for six years and return only to Kuwae once this period is up."

Marakipule did not ask any questions and he knew he had to listen to Pae and that something beyond his comprehension was going on. Marakipule's whole family took their places on Warasole and left in the direction of Aken. Pae then asked his brothers to organize a feast.

"Ha! He's organizing a feast when he's slept with his mother!"

"Laugh all you want! You won't be laughing soon!" Pae thought to himself.

A pig was sacrificed and Pae took its bladder, blew it up and hung it from the top of an ironwood tree. The next day, a pig was again killed and the feast continued. Pae took the bladder from the second pig and hung it under the first. The third day, another pig was killed and Pae hung its bladder under the first two. He repeated these actions until six bladders were strung on the tree. Six pigs had been sacrificed and eaten and their bladders were hung from the tree at differing heights.

As soon as the feast ended, Pae said "Now you will see!"

He climbed up the tree and popped the first bladder. The ground started to shake.

"What's happening?" "It's an earthquake!"

He climbed down a little and popped the second bladder. The ground shook again but stronger this time.

"Oh! What's going to happen? The ground won't stop shaking!"

"This is strange," some of the elders said, "the ground hasn't shaken like that for some time."

Then Pae burst the third bladder and the ground shook even harder.

"What is going on? The ground won't stop quaking!"

Others said, "What is Pae doing, bursting those bladders at the top of the tree? Don't forget that you slept with your mother. Stop showing off!"

"He's bringing us bad luck by acting like this. The ground won't stop moving."

Pae, not listening to the others, popped the fourth bladder and the ground shook, the land started to tilt and caused the people of Kuwae to be very afraid. Women and children from all around started to run into the bush. The men tried to stop them but they were too afraid. The elders threw themselves to the ground.

"Pae! Stop! Take pity on us and stop!"

But at that moment, Pae burst that fifth bladder and the island started to explode.

"Fire! Fire!"

"Lava! The volcano!"

Pae had just enough time to burst the sixth bladder.

"Your time has come! And my time as well!"

A volcano opened at the base of the tree. Lava covered everything and the volcano spat glowing blocks of stone. Pae's head was thrown all the way to Ambrym, and Kuwae broke loose. A good half of the island slid into the sea and Kuwae was destroyed.

At Maniura, the point between Eton and Forari, Marakipule had watched the explosion of Kuwae. The ground had shaken terribly on Efate and a tidal wave had followed.

And a tidal wave followed . . . hmmmm.

So we have a tsunami—probably generated by the eruption. The timing of this event also fits rather well with other oral recordings we have from elsewhere in the Pacific that point toward a much larger, region-wide tsunami that came from near Tonga. It is possible that this story has simply merged the two events, though this seems unlikely since we have geological evidence for both. The geological evidence for this Kuwae eruption and tsunami are pretty cool. The eruption left a huge hole in the seafloor—what had once been an island sitting above the waves was now a deep hole in the seafloor—all gone, and on the island right next door—Tongoa—is evidence of the pumice this eruption produced. In one spot 30 m above sea level, there is a lovely layer of wave sorted pumice with a huge, carbonized tree trunk rafted in on top of it, all put there by the tsunami that followed the eruption. Pretty cool (Figure 3.2).

This local Kuwae tsunami may be a slight tangent from the bigger event, but it is linked with a massive volcanic eruption that we know about, which dates to around the same time as the region-wide earthquake-generated tsunami as well, and it may therefore be linked with this larger event since it could have occurred during a period of increased tectonic activity in the southwest Pacific.

To put this in context, the recent (2022) eruption of the Hunga Tonga–Hunga Ha'apai volcano, which generated a tsunami that devastated Tonga and caused Pacific-wide damage (e.g., US$6.5 million damage in Santa Cruz harbor, California), was a minor blip in comparison. The 2022 event is causing scientists to scratch their heads a bit since the tsunami somehow managed to be Pacific-wide, and that surprised a lot of people. This recent eruption has been given a Volcanic Explosivity Index (VEI) of around 5, or Cataclysmic, as opposed to Krakatau (Krakatoa), a well-known catastrophic Indonesian eruption and tsunami in AD 1883 which had a VEI of 6 or Collosal. These pale into relative insignificance when stacked up against Kuwae with a VEI of 7, or Super-Collosal (I sense a shortage of superlatives here). As those of us who

Figure 3.2 Tongoa, Vanuatu—Carbonized tree (outlined by dashed white line) has been rafted inland on a bed of wave-sorted pumice (lower limit defined by black dashed line). The tsunami deposit is about 30 cm (1 ft) thick. (Photo: J. Goff)

work in prehistory know only too well, if you thought Hunga Tonga–Hunga Ha'apai was big (Plate 3.1), then buckle up because you ain't seen nothing yet!

For now, though, let us drift some 1500 km (over 900 miles) east to the island of Futuna in the Territory of the Wallis and Futuna Islands. This little-known archipelago is situated between Tuvalu to the northwest, Fiji to the southwest, Tonga to the southeast, Samoa to the east, and Tokelau to the northeast. It is actually French, or more correctly put it is a French over-seas collectivity (*collectivité d'outre-mer*). You should not be surprised if you have never heard of it, the entire land area is about 140 km² (55 sq mi), around three times the size of Manhattan but with only 0.5% of Manhattan's population. People have been living on the islands for nearly 3000 years, with Europeans only "discovering" them in the 17th century. Like many Polynesians islands they are rich in oral recordings, and there is one in particular that is of interest to us.

The Legend of Fatuloli refers to a place on the island of Futuna that was littered with coral boulders as recently as 1993. Sadly, they have all been destroyed and broken up over the years since then to use as building material for new houses and roads. However, the legend refers to their origins and the wishes of an old lady from a place called Sikupae.

"Ah, I wish I could have a pebble like those that are useless to the people of Vaika, because I need one to hold down the sleeping mat I'm braiding."

Three days she waited for her wish to be granted; at Tavai, at that time, there was only sand. A rock, coming north from the island of Alofi and passing offshore, heard her wish. This is why all of a sudden the sea rose up and a wave swept over the coast of Tavai carrying a vast quantity coral blocks. Not having had time to flee, the people were all buried. All the people perished, and when the wave receded, Tavai was covered with blocks of coral.

That's why we call this place Fatuloli (changed to stone?).

Here we have another catastrophic wave. Not surprisingly, anthropologists were unable to find anything useful to date it by though and so, based largely upon genealogy, they figured that this event happened about 400–500 years ago—a pretty shrewd guesstimate as it turns out. Some 20–30 years later, a visit by geologists found evidence (deposits) for two large prehistoric tsunamis, not just one, with the more recent of the two dated to around AD 1480 give or take a bit—perfect (the earlier one was about 2000 years old). This more recent deposit sits on top of what archaeologists call an occupation layer. In reality, to say it sits on top is a little simplistic. When the tsunami came onshore it ripped through the old coastal occupation taking lots of it with it and then dumping a bunch of pebbles, sand, and shells on top of what remained. What the geologist found was the non-bouldery part of the tsunami deposit that had not been taken away to use as building material for modern houses.

OK, that's two nasty tsunamis in the 15th century reported in the southwest Pacific, the one from the Kuwae eruption and the other from a big earthquake near Tonga, that occurred around about the same time, but these reports are from islands over 1500 km (more than 900 miles) apart.

What else?

Pukapuka, formerly Danger Island, is a coral atoll in the northern Cook Islands. It is one of most remote of the Cook Islands, situated about 1100 km (700 miles) northwest of the capital, Rarotonga (and about 1400 km/900 miles east of Futuna). It was probably first inhabited around AD 1300, and the original population is thought to have been as great as 1000 or more, but has recovered from extremely low numbers more than once. The most significant recovery came after an event that the people of Pukapuka refer to as The Great Death. The people speak of a night about 400 years ago as "*te mate wolo*" (the great death) when a giant wave washed over the island: "Waters raged on the reefs, the sea was constantly rising, the tree tops were bending low. On the next day all the island was broken and everything destroyed."

Only two men and seventeen women "with remnants of their families" survived to repopulate the atoll.

Now it is fair to say that there is some debate as to whether this was a large cyclone or a tsunami, but there is also geological evidence from some of the other Cook Islands for a tsunami at this time, so the jury is still out on this one. Since Pukapuka is a tough place to get to in a hurry, this is likely to remain a mystery for a while longer, but given what has been found on the other islands in the group, I favor a tsunami.

So let's just see what we might find elsewhere.

The North Island of New Zealand lies 3500 km to the south, yes, the South Pacific is a large place. Here a lot of research has been carried out on Māori oral recordings. We have the tale of Nuku-tawhiti who successfully called a mountainous wave ashore to rescue a whale, or of Manaia who attacked Nga toro rangi and his sisters. With canoes sitting offshore in the evening waiting to attack, Nga toro rangi suggested to Manaia that he attack him in the morning because among other things Nga toro rangi couldn't be killed at night, so it would be pointless trying. During the night though Nga toro rangi and his sisters uttered spells and created a large wave that came in and destroyed all of Manaia's fleet—everyone in the canoes died.

But perhaps the most well-known oral recording is that of Te Tai a Ruatapu—the Wave of Ruatapu, sometimes referred to as the Curse of Ruatapu (this oral record was remembered by local Māori when warnings were issued for the 1960 tsunami that came from southern Chile, interestingly). There are a few permutations and combinations that have been written down about this oral recording but the gist of it states that Ruatapu was angered by his father Uenuku who referred to him as "unchiefly." In revenge Ruatapu took a canoe full of all the sons of highborn chiefs out of sight of land and drowned them. He saved one, his half-brother Paikea, who swam back to shore after Ruatapu had told him that on a particular day he would send a large wave to drown everybody. When this happened Paikea and his family and friends survived by seeking refuge in the hills, while everybody else drowned.

These are very, very brief summaries of just three of a multitude of oral recordings from around the northern half of the North Island of New Zealand that speak of a large wave inundating the land. There is now so much geological and archaeological evidence for an event dated to between AD 1450 and AD 1480 (remember that the tsunami on Futuna was dated to around AD 1480 as well) that it forms the core of the recently established New Zealand Paleotsunami Database. Not bad.

Slipping slightly out of the oral records and into history for a second, some notable myopia has pervaded the written record since Europeans first arrived in the Pacific. This is not a dig at any particular group of scientists but rather at all of us because it is somewhat amusing to realize that once you know this next point, you see it reflected throughout the Pacific Islands as a testament to the truth and power behind oral recordings, extant or otherwise.

The devastating effects of the mid-15th century event that these oral recordings are referring to are particularly noticeable in the marked changes in settlement patterns that saw island populations move inland and uphill to get away from the coast. When the early missionaries arrived in Polynesia from about 17th/18th century onward they generally set up their missions at the coast on the highest ground around an island's periphery—what we would today call the storm ridge that was left behind by the largest storm. At the time of arrival, local inhabitants were living in groups inland and uphill, where they had moved and stayed after the mid 15th century tsunami struck. However, Polynesians were encouraged back to the coast by the missionaries who were keen to have their "noble savages" within easy reach of the missionaries (and future tsunamis). At the same time Polynesians were discouraged from using their native languages or communicating in anything other than the language of the new arrivals, and so sadly it was at this time that many of the oral recordings were lost. We will never know what is missing, or how much is missing, but what we can do is honor and value what is left behind and hope that they will help us survive the next tsunami when it happens, because on most Pacific Islands people are now living in the wrong place!

It is only now that the tide is turning, and it is our turn to learn from the fragments of a rich oral history that we almost destroyed. We have seen the memories of past tsunamis preserved not only in those of indigenous peoples but also in the ground. These two lines of evidence are like long lost siblings finally meeting up again after hundreds of years apart in the wilderness. But there are other siblings that are equally obvious once you know what to look for and yet equally hidden when you do not.

We need to learn about these.

4

Strand 3

New Light through Old Windows

Tsunami inundation . . . would cause multiple breaching of dune systems . . . assemblages would include remnant dune ridges, or pedestals, between each breach, and the individual overwash fans would coalesce to form sand sheets that may or may not be mobile depending upon aeolian and dune swale conditions.

J. Goff, D. M. Hicks, and H. Hurren. Tsunami geomorphology in New Zealand.

Tsunami research is a really new science in the big picture of things and there is a saying that goes something along the lines of "I keep seeing new things because I am wearing new glasses today." For those of us fortunate enough to not wear any corrective glasses or contact lenses this saying may seem a little strange. It has nothing to do with one-upmanship and the recent purchase of some new designer sunglasses that are quite frankly no better than the ones you can buy for a fraction of the price down the road at your local corner store. This saying relates to the fact that we keep learning new things about tsunamis.

While there are undoubtedly many researchers who focus in on their work like those Caesar Wegmann alluded to in the continental drift fiasco, there are equally just as many who reach out into the wider science community in search of novel ways to help in the study of past tsunamis. Often this can result in some quite remarkable findings.

Let us take a quick detour back to the Dog's Breakfast site. That deposit was sitting in a shallow, muddy lagoon and had been thrown in there by a tsunami that came rampaging onto the land and across into the lagoon. What we didn't say was that there had also been another more recent tsunami there.

In Search of Ancient Tsunamis. James Goff, Oxford University Press. © Oxford University Press 2023.
DOI: 10.1093/oso/9780197675984.003.0004

This was an exciting find because we now know that it was the AD 1826 tsunami, which occurred just as European explorers were arriving along the West Coast of New Zealand's South Island. There were some hints in the old records of sealers working in Fiordland, a little to the south of the lagoon, that a large earthquake and tsunami had indeed occurred then. It therefore became the default choice as a candidate for this younger tsunami, but it needed to be verified. However, the sealers' records referred to an area over 300 km (180 miles) to the south. So linking a tsunami in Dog's Breakfast territory with what the sealers saw would need some serious sleuthing.

The first European to travel down the West Coast and comprehensively explore the remote area near the lagoon was Thomas Brunner, and in late October AD 1847, 21 years after the AD 1826 event, he noted in his diary his observation of "the wreck of a large sealing boat amongst a lot of underbrush. It is about a quarter of a mile from high water, and the growth of the bushes and the appearance of the wreck show that the sea is fast receding from this coast." At another location, not far away, he observed: "The timber here is very small, and appears of recent growth. I think to the foot of the mountain range has been recently washed by the ocean."

These observations, made only a few days apart, give two different interpretations of what the small, scrubby vegetation might mean. Was the coastline receding or had the sea washed over it and destroyed the vegetation before the coastline returned to business as usual? What was going on? And what about the boat?

As most people know, stranded vessels are some of the more iconic pieces of debris left behind by tsunamis. Prior to the 2004 Indian Ocean tsunami, though, the most famous stranded vessels were probably the *USS Wateree*, marooned on land by the AD 1868 Arica tsunami in southern Peru (now northern Chile), and the steamship *Berouw*, carried inland by the AD 1883 Krakatau tsunami in Indonesia.

Like most evidence for past tsunamis though, time has not been kind to them. Once it was finally abandoned by the US Navy, the *USS Wateree* was used first as a hospital, later as a hotel, and finally for target practice during the Guerra del Pacífico (War of the Pacific, AD 1879–1883). It suffered from scavenging, souvenir collection, and most recently, relocation of the last remaining piece, a boiler, closer to the coastal highway as a tourist attraction. Ironically, scavenging and souvenir collection was aided by the AD 1877 Iquique (Chile) tsunami that broke the stranded vessel into pieces and moved them all back closer to the sea. The steamship *Berouw* on the other hand was carried 2.5 km (1.5 miles) inland by the tsunami generated by Krakatau's cat-astrophic eruption. Rusting iron relics of the ship could still be found well in to the 1980s.

The 2004 Indian Ocean tsunami, that we are all far more familiar with, had its own iconic ship, the PLTD Apung 1. This was a power-generating ship that was stranded 3 km (2 miles) inland in the city of Banda Aceh, Indonesia. It has now become both a tourist attraction and a memorial to those killed by the tsunami. The more recent 2011 Japan tsunami also had its fair share of stranded ships, although perhaps the most well-known was the 60 m long *No. 18 Kyotoku-maru* that was swept 500 m (300 yards) in-land into a residential district of Kesennuma city, where it was then caught in the fire that swept through the area in the immediate aftermath of the tsunami. The original plan was to keep it as a memorial of that terrible day, but it was decided to scrap it after an opinion poll found that nearly 70% of local people wanted it gone. While the wreck drew many visitors who prayed and left flowers at the site, this stark reminder of the tragedy was simply too confronting.

But what if you were to find a stranded vessel in the middle of nowhere? Could it have been put there by a yet to be discovered tsunami? This is some-what reminiscent of the saying "If a tree falls in a forest and no one is around to hear it, does it make a sound?"

Brunner's AD 1847 discovery of that stranded vessel on the West Coast of New Zealand was followed later in AD 1866 and AD 1867 by another dis-covery a little farther south. A Captain Turnbull reported the wreckage of a 600–700 ton vessel about 500 m (1600 ft) inland that appeared to have been there for many years. He was of the opinion that it was most probably one of the numerous whaling ships that frequented that part of the coast.

Hmmm—curiouser and curiouser.

More sleuthing was required.

After digging through a variety of archives, conducting a number of in-ternet searches, and wandering along the usual blind alleys, it transpired that there had been no big storms (or tsunamis) in the intervening years since the AD 1826 tsunami. As a result the coastal vegetation had managed to recolo-nize and start growing again undisturbed by any more catastrophic events. So what both Brunner and Turnbull saw was the natural regrowth of the coastal vegetation that had taken place since the vessels were stranded. There had also been no retreat of the sea as first conjectured by Brunner, no lowering of the sea level or conversely rising of the land, and so these strandings were indeed caused by the AD 1826 tsunami that threw the ships inland. We have no idea what happened to their crews.

What has all this got to do with the Dog's Breakfast site? A good question. Geologically there was a significant difference between the earlier Dog's Breakfast tsunami and this more recent event. There were no pebbles, no intact shells, and no wonderfully obvious tsunami deposit. The younger event was so subtle that it could easily have been missed. There was no obvious change in the geology, but when the muddy sediment was viewed under a microscope things became a lot more interesting. One thin, 1 cm (0.5 inch) thick layer of nondescript mud was chock full of marine microfossils. Fair enough, this is a coastal lagoon that is gently flushed by the tide every day, and so we would expect a lot of marine bits and pieces to also ebb and flow. But the lagoon is also threaded with the arterial channels of streams that empty into it from the glorious native forests that surround it and with them comes a constant supply of freshwater material. The interplay between these two diametrically opposed environments is reflected in subtle microscopic changes in the muddy layers.

A seemingly endless variety of microscopic beasties appear under the lens of the inquisitive researcher, but some are more useful and easier to distinguish than others, and these can say a lot about what was going on at the time they were deposited in the mud.

Here I feel that a brief microfossil detour is needed before we return to the story of new glasses.

Most people know about pollen—we see clouds of it occasionally in the summer and let us not forget about hay fever and the important vagaries of the pollen count for sufferers. Technically, it is grass pollen that causes hay fever, although to be honest anybody with an allergy to the large quantities of lightweight pollen produced by anemophilous or wind pollinated plants undoubtedly calls their symptoms hay fever. Pollen can travel great distances and is easily inhaled into nasal passages. The resulting allergies in those who are sensitive are by no means trivial and are known to debilitate well over 20 million people in the United States alone.

Palynology is the study of pollen, and the abundance and variety of pollen preserved in our mud reveals important information about past climates and the vegetation in the surrounding landscape. You can look back over thousands of years by studying mud samples from deeper and deeper below the ground. But these microscopic grains only come from the land, they do not come from the sea. In other words, if you find a lot of pollen grains in what you think is a tsunami deposit then your interpretation is probably wrong, and you are looking at some type of river flood. It is literally almost as simply as that!

What we are interested in are diatoms. These will be far less familiar to most people, but if you talk to someone who studies them you will find out that they are much prettier and more interesting to study than dull old pollen grains . . . in my opinion. So what are they? They are microscopic, single-celled algae that can either be found as single solitary cells or in colonies that can take the shape of ribbons, fans, zigzags, or stars. Individual cells range in size from 2 to 200 micrometers/microns. Living diatoms can be found pretty much everywhere, in the oceans, rivers, lakes, and soils of the world and are responsible for generating between 20–50% of the oxygen produced on the planet each year. So they are both very small and very useful. One more fascinating factoid about these incredibly pretty little beasties is that they live in glass houses—their minute little shells are composed of silica! They also live in almost any type of environment you can think of—they float in water like plankton, they attach themselves to plants, debris, or other critters, and they even live in the soil and sediment. So if you combine the facts that they have nice hard shells, live (and die) almost anywhere, and look very pretty under a microscope lens, then you can no doubt imagine that scientists love studying them for many reasons. There are a huge number of different types of diatoms, and so changes in the type and their abundance can tell us a lot about what the environment looked like at the time they were alive.

One word of caution to those who are suddenly bitten by the desire to see some of these "in the flesh," so to speak. It is probably best to look at them online and see beautiful examples of wonderfully preserved diatoms, the photogenic ones. The reality is invariably not quite as sexy. Do not rush out and buy a microscope and smear a bit of mud on a slide in the hope of seeing some wonderful diatoms. It takes a bit of work to isolate the diatoms from a soil sample, and just to make matters worse, it requires a different process to isolate pollen grains from a soil sample too, so there is a lot of work involved if you want to look at both of these microscopic remains. For diatoms you have to get rid of any bits of soil, any bits of organic material like leaves and so on, and then any other microscopic beasties that might get in the way, and so on. The process is somewhat tedious, requiring assorted chemical soaks and seemingly endless centrifuging of each sample and you can end up with . . . nothing. Having said that, you can also end up with some really cool results, but it takes time and the correct laboratory facilities, the correct training, and a degree or two to help you on your way. After all that time spent learning about them though, you will still be amazed by how pretty they can look.

In our little study of the world in the sediments of a gorgeous coastal lagoon these variations are invaluable. As we look at the casings or frustules of dead diatoms preserved in the mud taken from deeper and deeper down in a hole in the ground we see before us a story of the interplay between the forces of the land and the forces of the sea. Big river floods bring down copious amounts of freshwater diatoms, such as the elongated and pennate (bilaterally symmetrical) *Pinnularia* species. While some *Pinnularia* can be found in estuaries and the sea, most—and these are easily identifiable to those in the know—are freshwater algae, usually found in ponds and moist soil. Conversely, when the sea catastrophically inundates the lagoon, during a tsunami for example, in rush innumerable marine species such as the centric (radially symmetrical) *Paralia sulcata* that is commonly found in both planktonic (floating in the sea) and benthic (on, in, or near the seabed) environments.

And in this one thin layer of non-descript mud, no more than a centimeter thick, was a veritable treasure trove of marine diatoms. Yes, *Paralia sulcata* were there but then so were many more, *Cerataulus turgidus*, *Actinocyclus octonarius* var. *crassus*, *Grammatophora oceanica*, *Triceratium* spp. and *Navicula lyra*, just to name a few, and sounding like a roll call of long dead Roman emperors. Here was evidence of a big wave of seawater rushing into the lagoon, but where was the sediment? Where was the smoking gun that said to us that something big and nasty had rampaged into this lagoon? Was this all the evidence to be found for the AD 1826 tsunami?

The answer to that is both yes and no.

Yes, that is all the physical evidence we have ever found for the AD 1826 tsunami deposit. Yes, it was actually a much smaller event than at the Dog's Breakfast, but it was still large enough to flatten all of the vegetation, killing it with saltwater, and causing Brunner, 21 years later, to observe that the vegetation seemed to have only recently grown back after being washed by the sea. Tsunamis do not have to be massive to do such damage, they do not necessarily carry much sediment with them either, it rather depends on what is around to be moved, but also it can depend upon what caused the tsunami in the first place.

Here there was a clue, a clue that only new glasses revealed. In the early days of tsunami research it was all about the sediment—the visible evidence of a nasty event, and then came the more clever stuff, the invisible evidence, or at least that only visible using microscopes and other devices. So when we found the AD 1826 tsunami by looking at the microfossils we were doing what we thought was the clever stuff, we had found something that no one had seen before. Since then, though, this has become one of the more commonplace ways of looking for evidence. There have been several new pairs of glasses since then!

After a few years, a few dots started to join up. We had the first two dots in the AD 1826 tsunami mystery. First, there was no obvious change in the sedimentary record that we could find. Second, there was a truck load of marine microfossils in the mud. The additional dots were obvious, really, but were part of a much bigger picture story that didn't factor in during the early days of tsunami hunting. First, the mechanism that caused the tsunami must have somehow been related to the large earthquake recorded over 300 km (180 miles) to the south. But at the Dog's Breakfast site, the fault line most likely responsible for the earthquake was several kilometers on the landward side of the lagoon, not out at sea where it could generate a tsunami. To have generated this tsunami though there must have been some displacement of the water in the sea not far from the lagoon—the sea must have been upset enough that it produced a tsunami. But as the source of the earthquake was on land, it could not have caused the tsunami, unless the earthquake groundshaking caused a big landslide to fall into the sea nearby, or something similar. But there was no evidence for that.

Hmmmmm—where on earth could this tsunami have come from then?

The answer to this has become more and more apparent and obvious as the years have gone by, and our glasses have improved, but this was the first building block in the monument of evidence that now stands before us.

Ōkārito Lagoon is surrounded by a beautiful pristine forest that comes right down to the shore in many places. If you choose to paddle out in a kayak or take a more leisurely cruise in some motorized craft and stop in the middle of lagoon and look around you initially see trees, an endless sea of them. There are rimu (*Dacrydium cupressinum*), kahikatea (*Dacrycarpus dacrydioides*), and kamahi (*Weinmannia racemosa*), with an understorey of lancewood (*Pseudopanax crassifolius*) and cutty grass (*Gahnia rigida*). The latter are the bane of the intrepid forest explorer—a tall plant with long grass-like leaves that are serrated and sharp on both edges and will inflict cuts, often quite deep, if they are held (grabbed in a panic to stop falling in a hole hidden by the grass) or brushed past.

A closer inspection though reveals a little terrace 50–100 cm (1.5–3 ft) above the water level that encircles the lagoon, above which the land rises up as either a small cliff or a steady incline into the hinterland. The trees are noticeably smaller and more spindly on this lagoon-side terrace. Strange. Is it washed by the sea, making life difficult for the trees to grow? Unlikely, since it is only around the lagoon that we find it, and the sea can't get there. Is it too close to the water for trees to grow properly? Possibly. Or is it simply that they are a little younger?

Dendrochronology (or tree-ring dating) is a way of working out to the exact year when a tree started growing—we have come across it earlier. New growth in trees occurs in a layer of cells near the bark, and the rate at which these cells are added varies throughout the year as the seasons change, with lots of growth in the summer and a thinner, denser area where less growth occurs in the winter. Put these two bits together and you have an annual growth ring. Count the rings and you have the age of the tree. The great news is that you don't have to cut the tree down and kill it to find out how old it is, you can use incremental borers more commonly known as tree corers. These are like hollow tubed screws that are literally screwed into the tree and when they are pulled out retain a small core that has cut through each tree ring to its center—plug the hole to protect the tree and go and count the rings. A simple process that can reveal so much.

Results from the lagoon edge were compelling. The older trees inland from the terrace dated to between about AD 1600 and AD 1800, but those on the terrace were no older than AD 1832 (Figure 4.1). Bearing in mind that it takes a few years for seedlings to get going, these results showed that the terrace formed . . . in AD 1826, the date of the earthquake and tsunami! Ah ha— a light bulb moment, a cleaning of glasses, and a mystery solved, or at least a mystery solved for this lagoon. Later on it was solved for an entire length of coastline several hundreds of kilometers long, but that took another 15 years.

The mystery of the tsunami was clear. In a sense it was generated by the earthquake hundreds of kilometers to the south and yet in another way it was not. What seems to have happened is that the tsunami experienced by the sealers was caused by a fault movement underwater—no problem there. But here, to the north of where the sealers were, the fault was under the land and could not generate a tsunami. BUT, and this is very important, what we do know is that when this fault—the Alpine Fault—ruptures in a large earthquake, this land-based part of the fault does three things. First, on the landward side of the fault (east) the land goes up helping to build the magnificent

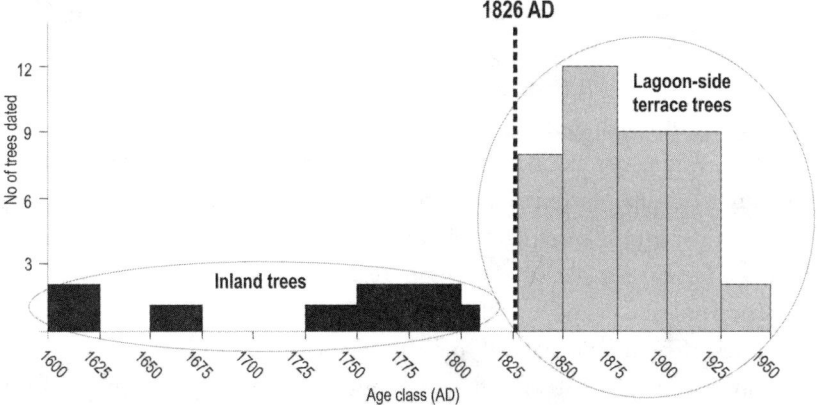

Figure 4.1 Lagoon tree ages: Inland the youngest dates to AD 1811, on the lagoon-side terrace the oldest dates to AD 1832, showing that the terrace formed after the AD 1826 earthquake and tsunami. (Image: J. Goff)

Southern Alps over the millennia. Second, the land on the seaward side of the fault goes sideways, sliding horizontally, a lot. These two useful bits of information have been known about for decades. The complication in this case though is that the AD 1826 earthquake was not massive, and it also really only took place on the southern part of the Alpine Fault (or maybe an associated fault), not all along its length, and not up in the north near to our site . . . hmmm.

The third point is the kicker, and over the ensuing 15 years we have found more and more evidence for this up and down the coast, even though the full story is still far from clear. There appears to be a parallel fault somewhere between the sea and the Alpine Fault. Every time a big one goes off on the Alpine Fault this other fault seems to behave rather strangely—it either drops down, or subsides, by a matter of tens of centimeters (a few inches), or uplifts by a bit more than that, and it seems to alternate this movement with each of the big earthquakes. In AD 1826 it responded to the fault movement hundreds of kilometers to the south by uplifting by about 50–100 cm (20–40 inches), creating a small terrace around the lagoon that sat there waiting for seeds to fall on it and germinate. This uplifting of the entire lagoon pushed the water out into the sea, only for the sea to return the favor, and a local tsunami rushed in to fill the recently emptied hole.

Interestingly, in the previous large Alpine Fault rupture the land subsided by about 50 cm (20 inches). The consequences of this somewhat bizarre response are seen far more clearly in coastal sediments to the north and south, where there are both uplifted terraces and drowned forests of various ages all intermixed with each other, as the magnitude of each yo-yoing of the land varies with each event.

While all of this rather neat science helps to prove the date of the AD 1826 tsunami, it has also started to raise a few worrying questions that are yet to be fully resolved, and also helped explain a few things too. First of all, every time there is a big earthquake in the region it causes a vast number of landslides as the ground shakes and unstable slopes collapse. This in turn generates vast quantities of sediment that fall into the rivers and is gradually transported downstream to the coast, or in this case, into the lagoon. Now, if the land has subsided then there is lots of new space, a deeper lagoon, for the sediment to settle in, but if the land has uplifted (as in AD 1826) then a shallower lagoon means that there is less room for it to settle out in. But it is still a lagoon, and it continues to fill in until presumably the next earthquake causes it to subside and create a deeper lagoon (what we would call more accommodation space) with more room for sediment to accumulate.

That is all logical.

OK. So, now the lagoon is, not surprisingly, very shallow because lots of sediment is being deposited in what became a shallower lagoon in AD 1826. The lagoon has accumulated at least a meter of sediment since that earthquake and will effectively fill in within another 100 years at the latest. BUT a study of deep sediment cores taken out of the lagoon shows that this has never happened over the past 6500 years or so—it has never completely filled in. In other words, the next big earthquake is incredibly likely to happen within the next 100 years. It is also quite possible that the land will subside next time, AND this movement will do two things—generate a tsunami as the water rushes in to fill the new space created by the land going down, and quite possibly bring a lot of buildings along the coast just that little bit closer to sea level . . . gulp. Well, that is how I see it at the moment anyway, but you never know, new glasses may "see" new things.

Let us now leave the Dog's Breakfast site and its younger cousin behind and take our new glasses out for a spin. We have certainly learned a lot from that place though, which should stand us in good stead for our next mission.

One of the great things about the global tsunami community is that we all come from different backgrounds with different skills and training, and on many occasions this has led to some quite remarkable innovations, some of which will become apparent as we delve deeper into the topic. But for now, let's stay fairly close to the topics we have talked about so far.

The most common signature of a past tsunami is a thin layer of sand (yes, there can be anything from boulders to mud or even less) that generally starts off thicker at the coast and gets thinner as it goes farther inland. This is a common feature all around the world. Also, because a lot of scientists have spent a lot of time studying microfossils in tsunami deposits, we know that a lot of the material comes from the seabed near the shore but that an awful lot is also picked up from the land and the two are mixed in with each other.

Excellent, A+, well done.

It therefore seems strange that only a very few scientists have looked at what all of this action did to the landscape, its shape and form.

Many years ago, while I held a brief dream of being able to function as a reasonable amateur artist, I remember my teacher saying to me that, "Negative space is visually just as important as positive space—it's the area around and in between your subject matter." That was a concept that struck my brain as being extremely logical, but what also struck my brain was my complete inability to produce that on paper. My putative artistic career was not immediately crushed, I did manage a single watercolor painting—abysmal—and while my one oil painting was quite good (in my mind), I only managed to convince myself of its worth by continuing to pile oil paint on top of the mistakes, and so it ended up as much a sculpture as a painting! However, it was my pen and ink portfolio of predominantly rugby cartoons that was (and is) my pride and joy. They almost made it into a calendar, but I backed out at the very last minute and have never quite had the guts to revisit that enterprise. I did sell a grand total of four of them, so for those who own them—you have a treasured family heirloom that in years to come may end up being one of those incredibly obscure pictures that comes up on Antiques Roadshow and baffles the expert who will have never heard of me. One can but hope it holds its value.

The point was well made by the artist though, and here once more we can see the polymathematical world of tsunami research. Art has its place in this ragtag band, so perhaps I should not be saying tsunami scientists but rather tsunami scholars. But I am sure that I will slip from this excellent suggestion as I fall back into the warm embrace of my science family—no slur intended to the numerous excellent artistic contributions.

To achieve a balanced canvas the artist must focus on both the negative and positive space. Equally, to achieve a balanced understanding of tsunamis, we must try and come to grips with this negative space as well. This is easier said than done. Just like a completed canvas, environmental and physical processes set to work in a natural aging process. For a painting this is generally regarded as acceptable and can even enhance the value. For preserving the evidence of a past tsunami, these processes are a nightmare. The instant a tsunami has laid down its deposit, things start to change it. The sediments laid down by the "run-up" of the water onto the land will immediately suffer a degree of erosion and reshuffling when the water rushes back out again in the backwash. Great start. And so it goes on.

But in many cases, the negative space left behind by erosion and reshuffling—by both the run-up as it interacts for the first time with the land and the backwash as it finds its way back over the newly created tsunami landscape—can survive for much longer. After all, it's not as if another really nasty event like a tsunami is going to happen for a long time, hopefully. And so this altered landscape can sit there waiting to be discovered. This ability to survive longer *in* and *as* the landscape depends to a large extent upon what normal physical and environmental processes are operating in the neighborhood—just as for a painting.

The relationship between negative and positive space is obvious in many ways. But a focus on the positive is easier because it has so obviously been added to the landscape whereas negative space is harder to study simply because it isn't there—even though it is.

A tsunami sand sheet is largely draped over the landscape. At some point during this draping process the landscape may have been changed by erosion, before something was draped over it soon after. This coastal drape can cover a range of surfaces—a flat beach, a sandy plain, a marsh, a lake bottom, roads, gardens, and so on. To get the vast amount of material necessary to drape an extensive layer of sand over the landscape, the tsunami has to pick it up, erode it, from somewhere. There is plenty of material out at sea. But this is a wave with a vast amount of energy, and plenty is left in the tank as it comes inland, and so it erodes.

The first and most obvious thing a tsunami often meets as it comes inland is sand dunes. As we have seen, these are absolutely wonderful natural barriers that help to protect us from much of the anger of the sea. They can take a pounding, be half eroded by brutal storms, and then gradually over the ensuing months build themselves up again in preparation for the next fight.

That is what they have done for millennia although it is worth noting that this beautiful example of the balanced equilibrium between land and sea is starting to change quite markedly as sea level rises.

Humans have done their bit to mess up this equilibrium between land and sea—we saw what simply removing sand dunes did for Sri Lanka in the 2004 Indian Ocean tsunami. But we have other ways of mucking them around that take a little more time to kick in. The simplest question to ask on this front is, where does the sand come from that makes up the sand dunes? The answer to that goes a long way to showing just how stupid we are.

As anyone who has been down to a sandy beach for a day out will know, if the wind blows then the sand goes with it, although some of it doubtless doesn't go too far, finding a resting place wherever it comes across an obstruction and is slowed down a bit—hair, teeth, shorts, bikinis, shoes, sandwiches, ice creams—essentially anywhere you don't want it to be. There is nothing more annoying than finding a nice, secluded spot in the sand dunes only to have someone walk by above you generously stirring up the sand and letting the wind do the rest, driving it into every nook and cranny. Yes, the sand comes from the sea. It is delivered there mostly by rivers and works its way along the coast by a simple process called longshore drift—water currents just offshore slowly shift the sand sideways until at some point it gets shunted up onto beaches by the waves. It is dried by the sun and wind and then, yippee, assuming it doesn't end up in your hair it joins its mates in the dunes and settles in for a break, only moving every now and again after that.

So what happens if, for example, a human who thinks they know best puts a dam on a river? Or what happens if they build a big breakwater out into the sea to protect their bit of coast and allow ships and boats to get from the shore out to the sea more easily? The sand supply is cut off, beaches get no more sand, dunes get no more sand, and every time some of the sand moves on in the wind or is taken away by a storm, it doesn't come back and the coastline starts to erode.

But enough about we humans, although our influence will become important again at some point later on.

Sand dunes, as we know from the 2004 Indian Ocean tsunami, can help to greatly reduce the damage caused by a tsunami by absorbing and reflecting much of the brutal energy of the waves back out to sea. In some cases though, even sand dunes cannot hold all that energy back. Sand dunes usually either sit at the coast as a seemingly endless "field" of hummocky land extending inland for hundreds or meters or more or form a long ridge parallel to the coast, pretty much one dune thick. Both are important to our tsunami story.

Imagine you are a tsunami. You have energy to burn as you come rampaging across the beach, and there are the dunes in front of you. You throw yourself at them, and while most of you gets pushed back, the dunes aren't all the same height. They don't form a solid wall, there are gaps, low points through which you can power through and continue your inexorable journey inland. As you power through this gap you pick up huge amounts of the sand, eroding the sides and bottom of the gap and carrying the sand inland with you.

Yup, it is really as simple as that. The technical process here is that a tsunami cuts a breach, or cleans out an existing one, through a sand dune or beach ridge. The material scoured from the breach usually gets deposited quite quickly and forms a large fan of sand, a bit like a delta pointing away from the sea, as the water emerges inland of the ridges. Just to complicate things even more, because the tsunami has just carved out this nice new gap, when the water reaches as far inland as it can go it then drains back out to the sea, often through these newly created gaps, picking up some of the fan it deposited on the way in and creating another fan on the ocean side of the dunes on the way out! These big fans of sand can be pretty thick. In Japan, after the 2011 tsunami, we studied one that was almost a meter thick. This is very useful because compared to the thin weedy little sheet the tsunami drapes over the landscape, these fans are far more likely to hang around in the landscape for centuries. So if you know what you are looking for, you can see the complete assemblage—the negative and the positive spaces—the tsunami breach (the gap it carved out) and the fan or fans associated with it. Since these can be pretty obvious features, you can even see them on aerial photographs or satellite images.

A really cool example of this comes from the ocean-wide tsunami generated by the giant Chilean earthquake of May 22, 1960. These scour and fan features were found near the city of Valdivia in Chile on aerial photos taken in January 1961. You can still see them today if you know where to look because there hasn't been another big event since then to reshuffle the landscape.

Now that is all well and good, but the real fun comes when you string a whole bunch of them together. So instead of one breach and one fan, you end up with one after another after another after another and so on. What does that look like? Well, we have seen such things from recent events such as the 2004 Indian Ocean tsunami, but in New Zealand researchers have found these features preserved in the coastal landscape at least 500 years after the tsunami struck. If you take any one of the scours in isolation, it looks just like their 1960 cousins, but string them all together and the assemblage of negative and positive spaces becomes truly fascinating . . . and once you know about it, it becomes so obvious that you start to wonder what all the fuss is about, the evidence of past tsunamis is easy to find. Hmmm—well, let us continue to divulge the tricks of the trade then.

Take one tsunami scour and cut and paste this along say a couple of kilometers of coast and you end up with a series of highs and lows. The highs stand there like isolated lumps sitting there all alone, stranded, reminiscent of the first lines of Shelley's poem, Ozymandias:

I met a traveller from an antique land,
Who said—"Two vast and trunkless legs of stone
Stand in the desert . . ."

We call these vast and trunkless legs of sand pedestals. In one of the interesting twists of tsunami geology, while individual tsunami breaches had been reported for several historical events, no one had ever gone beyond that and ever found a string of them all in a row. That honor was left for paleotsunami research, and then it became an obvious feature in their historical counterparts.

It wasn't a particularly stunning discovery, but it shows the importance of understanding context. When working in areas with sand dunes there are certain features that you see all the time, such as blowouts. These are depressions or hollows formed by wind erosion on a sand dune that tend to be formed together, with an inland accumulation of sand, a fan or lobe, where the blown out material is redeposited. Sound familiar? What is the difference between a blowout and tsunami breach? They look exactly the same, but the former is created by wind and the latter by water. To the trained eye they simply look different, and also, remember, a tsunami breach is just part of a whole complex of other deposited material.

What then if you are walking along the beach looking at the sand dunes and you see a whole series of these depressions separated by pedestals. How did they get there? While I would love to say something like, "well duh, blowouts tend to be few and far between, you don't get a whole bunch like this," the truth is, quite naturally, far more complicated than that. So, let's cut to the chase.

The first time we saw pedestals they were in a line parallel to the shore. There was no real regularity to the gaps between them, but they were all quite old and weathered. The depressions between them were wide and deep, so deep that while the sand near the top looked like it was windblown and unstable, lower down it got darker, more orange and weathered, more compact and eventually downright hard and coffee colored. These were not formed by the wind, no wind could have shifted that stuff, and then there were so many of them.

Sit back a second and just think of the logic here. We are on a beach, there is plenty of fresh sand around, but it is not filling in the gaps. That is because these gaps are now acting as wind tunnels and the sand blows through the gaps, but how did they get there in the first place? If the wind couldn't move all of the material, then these assemblages must be some relict feature of an event that was so extreme it rejigged the sand dunes so much that they have never been able to return to the state they were in before. The previous sand dune system that had been in a happy equilibrium with the wind and the sea, seeing a bit of give and take over the days, months, years, and decades, had been so upset by whatever had happened to it that it never properly recovered.

If you take wind out of the picture, then the only variable you have left is the sea. If it was the sea then it must have been something pretty brutal—either the mother of all storms or our dear old friend, a tsunami. Could it have been a storm? Well, you could wander down to somewhere on the southeastern seaboard of the United States and see some very impressive hurricane washover fans with little pedestals between them, but these fans do not go very far inland and these features don't last long because sand fills them in again. This is "normality," it does not flip the balance beyond what this type of coast faces over the years and decades. But of course there are other things to look for to also support the somewhat outlandish suggestion that these were caused by an ancient tsunami—you do the geological research inland just to prove that ultimately what you found without even needing to look at the geology was what we now call tsunami geomorphology.

Neat—but there are more than just pedestals in this little assemblage of negative and positive space. Inland from a single tsunami breach there is fan of redeposited material just like in a wind blowout, but inland from a series of pedestals the sediment rarely settles back down into a nice simple fan, because here the energy of the tsunami spreads the fans out much more, in an unusually thick sand sheet that then, yes, you've guessed it, thins inland as we would expect.

That all sounds simple, and we have in more recent times seen exactly that type of assemblage—pedestals, really thick sand that then starts to look much more like the drape-like, thin sheets of most tsunami sand deposits. Perhaps one of the coolest of these in recent times comes from near Kesunnuma, Japan, after the 2011 tsunami smashed through the dunes (Figure 4.2), among other things depositing a ship some 500 m inland. And so here is a case of new light being cast on the evidence for tsunamis by looking through the windows of older events. But the research goes further. What happens over time to the really thick deposits that were dumped immediately inland of the dunes?

Water and wind have their role to play here too, after all they are part of the normal state of affairs in this type of coastal environment. Since the poor old sand has only recently been relocated to this new inland spot it has no vegetation to bind it together, it has been put there by a somewhat unusual event

Figure 4.2 Tsunami pedestals near Kesunnuma, Japan. Note the two story building in the background destroyed by the tsunami. (Photo: J. Goff)

and so it is probably not "settled" where it is. This is borne out by the fact that there is also a big wind tunnel just downwind of it now that blows through the gap created in the dune. And so the wind blows, the sand ups sticks again and marches even further inland—the finer sand is suspended in the wind on its way inland but most of it is probably transported by saltation. Saltation, from the Latin *saltus*, meaning leap, is pretty much that, the sand, being too heavy for the wind to carry all the time, effectively bounces along the ground after briefly being carried for a short distance. Imagine you have just been to the supermarket and filled your bags with groceries. Easy enough to pick them up and put them in the back of your car, but back home you have to carry them a 100 yards to the front door. The light ones you can carry all the way (suspension), but the heavier ones are too heavy to do in one go and so depending upon how heavy they are you may have to stop several times (saltation) on your journey.

Eventually the sand settles inland at a happy equilibrium between the wind speed and the weight of the particles of sand, and this accumulation of sand forms a parabolic dune, a U-shaped mound like half a paperclip, pointing inland with its convex nose trailed by elongated arms. This pattern shows that the wind blows harder in the middle than at the sides.

OK, so we are starting to make some progress here. Closest to the coast we now have our pedestals and on the inland side we have a bunch of half paperclip shaped dunes. Cool, BUT we are not there yet, and it is around about this point that an almost deafening cacophony of sound from some potentially disgruntled colleagues starts to descend. Why?

Well, even I remember learning about this in school—parabolic dunes are formed from blowout dunes where the erosion of vegetated sand leads to a gap being carved out in the dunes by the coast and the sand being blown inland to form our paperclip dunes as the next row of dunes inland. All very logical, all correct, but just because that is the original way researchers figured out how these things form doesn't mean that it is the *only* way. This is where we enter into a rather fun, if that is the word, part of the science where you can almost sense the battles between different scientists.

To a large extent the study of tsunami geomorphology sits somewhere between those who work on tsunamis and recognize that these features exist and those less involved in tsunami research who are far less willing to accept the idea. This interesting state of affairs, where two processes can end up with the same result—in this case, parabolic dunes—is called equifinality. Many years ago a colleague and I decided to write a lengthy review paper about equifinality in landform formation. After a year or two of quietly assembling a seemingly endless collection of references and papers we finally decided that neither of us had anywhere near enough time to do justice to the topic—but it is a fun one if anyone wants to pick up the torch and run with it.

The simplest way to recognize tsunami geomorphology is context: (a) There are multiple pedestals and "blowouts" along the coast that all formed *at the same time*, not just one or two weak points where the wind managed to destroy the vegetation and create one or more blowouts over a period of time. (b) Given this blinding logic then there should be a group or field of similar-aged parabolic dunes that formed inland—not just one or two. (c) Finally, and invariably absent from your average wind generated blowout, between the two—pedestals and parabolic dunes—there is usually a large area of "hummocky topography." Ah ha—the missing link.

Here we have the obvious difference in the two processes AND another case of equifinality. First of all, when a single blowout happens it is caused by the wind and so once picked up the sand is blown inland and usually goes a fair way until it settles down. With tsunami "blowouts" there can be lots of them in a line and the sand is carried inland by water, usually being dumped just inland as a series of wet, delta-like deposits that all coalesce to form a vast linear expanse of upset sand. It doesn't move too far inland, and it has to dry out first before the wind can start blowing it around, and even then there is so much of it that a lot of it goes nowhere and just sits there doing nothing.

But wait, there's more. Mother Nature is nothing if not persistent. The wind can only do so much, and so we are back to water again, rain to be precise. The rain falls on this big pile of sand behind the dunes and runs off it, creating little rills and gullies and eventually an undulating and lumpy landscape that we call hummocky topography.

Again, all very logical and sorry for the long explanation. BUT hummocky topography is a bit of an amorphous blob of a phrase, after all, vast tracts of land could fall under that category. Throw in a few hills and there you go—hummocks. The phrase actually got its first outing into the world to describe the vast tracts of hummocks that are commonly found near the edges of the old ice sheets that formed during the last glaciation. Depending upon where you live the last glaciation finished somewhere between about 10,000 and 12,000 years ago as the power was turned off on the big deep freeze. The huge ice sheets that covered much of northern Europe and North America slowly collapsed. All the debris carried in, on, and beneath the ice was unceremoniously dumped creating, you've guessed it, a hummocky topography.

Yes, nowadays, hummocky topography pretty much looks the same whether it was deposited by ice or tsunamis, but that's the key—ice and tsunamis. Context. You rarely find coastal sand dunes next to glacial hummocky topography, and if you do, a quick dig of the hummocks and you will find them composed of a bewildering variety of stuff that had once been part of the ice's cargo, nothing like tsunami-transported dune sand. On yes, and no parabolic dunes.

Now that we have that out of the way—so what? This is where it gets even more interesting. In essence, what we have done here is assemble all of the key bits and pieces of a tsunami geomorphology. We have shown why this particular assemblage: pedestal-hummocky topography-parabolic dune cannot be created by wind and most definitely not by glaciers. If you wander through some dunes somewhere with the right glasses on then you may even be able to spot this assemblage by yourself, sometimes in small bays it may simply consist of three of four pedestals and so on, but in big bays it can be a much more extensive feature. But that all takes time and a bit of luck. A much better way to find this assemblage is to use LiDAR.

LiDAR is an acronym for Light Detection and Ranging, a bit of a play on words, or another acronym if you like—Radar, that stands Radio Detection And Ranging, although no one ever thinks of it as an acronym now. LiDAR, or LIDAR, LIDaR, Lidar, or lidar—there is no real consensus on this—had its first iteration in 1961 soon after the invention of the laser which is the "light" source this tool uses. The basic premise is that a laser is used to calculate the distance from the earth of individual points at a very high resolution. Fly an aircraft with a LiDAR set up on-board across a town and fire laser beams down to the earth. These then bounce back to a receiving device on the plane and the time it takes to bounce back is a measure of the distance—so it would take longer for the pavement signal to bounce back than say for the top of a building and so on. The precision is amazing, enough to find something just a few millimeters high. All the data points are then put together to produce an incredibly accurate 3D image of the earth's surface in the area you are interested in.

LiDAR was first used in 1961 to track satellites but needless to say its military uses were soon realized and it was used as a large rifle-like laser rangefinder for targeting. But it had bigger fish to fry. In the United States it was used in meteorology to measure clouds and pollution. Then in 1971 it was used by Apollo 15 to map the surface of the moon. It now has many uses, but in the earth science line of business it can, for example, map earthquake fault lines hidden by thick forest, showing details such as minor vertical differences on either side of the fault or even electricity poles tilted by a recent quake. No need to spend hours and days walking around trying to find this evidence—do the LiDAR survey first, and then use that to decide where to go on the ground for a closer look.

LiDAR can also be used to study the coastline. There's not necessarily a lot of trees here but there is lots of lumpy stuff—LiDAR's bread and butter.

So here is a scenario. In the southeast corner of the South Island of New Zealand there are several Māori oral recordings or *pūrākau*—as we know these are sometimes erroneously called myths or legends. A few refer to the same place and with the same basic information and are useful for cross-checking details. But a word of warning, the versions I have used here has been filtered through the lens of the Victorian ethnographer who wrote them down. They are by no means complete or necessarily correct. They come from around the Moeraki Peninsula (famous for the Moeraki Boulders—but that is a major tangent and all I will do is recommend finding out a bit about them, they are very impressive) and refer to Rakitauneke, a famous *tohuka* (priest—elsewhere in New Zealand this would be *tohunga*), who had a guardian whale called Tu-teraki-hua-noa. One day the whale appeared off Moeraki, and the children cursed it. In anger, Rakitauneke sent in a tidal wave, which drowned them. They were standing by the freshwater Ka-wa creek at the time, which as a result of the inundation has been brackish ever since. A variation of this *pūrākau* says that Rakitauneke brought up lots of fish for the people because they wanted food but they complained that there were too many. As a result of this he called up Ruatapu (remember the tsunami story from earlier) who sent a big sea on to the land and washed the fish off.

Could there have been a big tsunami that struck the coast there? There is a lot of extremely culturally important archaeological sites along that coastline, and you can't simply have a herd of geologists (or would that be a "seam of geologists" or maybe even a "sample of geologists"?) running roughshod through such sites, they are way too important. The safest, and in many cases the only option, would be to find out if there was any possible evidence for a tsunami without even touching the ground, and then try to work out which are the best, non-archaeological sites to search and take it from there.

Fortunately, much of New Zealand's coastline has been surveyed by airborne LiDAR, so there is a lot of data to play with. Since Moeraki Peninsula is largely made up of cliffs and small bays it is not the right type of environment to look for tsunami geomorphology, but to the south there are several good sandy bays. The data search began. The result was not a long time coming, bay after bay after bay had the same assemblage. Pedestals, hummocky topography, and where there was enough room for the sand to blow inland, there were parabolic dunes and hummocky topography as well. In one simple exercise tsunami geomorphology was born—big time. Armed with this information, further searches with a dash of geology thrown in revealed that a previously unknown tsunami had struck this coastline sometime in the 13th–15th centuries. Of particular note was an archaeological report from the early 1980s on a coastal site where the archaeologists quite clearly stated that the area showed evidence of having been "washed by the sea." Over 30 years later, a geological study of the surrounding area revealed an unusual layer of pebbles and (fining-up to) sand approximately 20 cm thick sitting on top of the sandy areas around the abandoned archaeological occupation site . . . hmmmm.

A previously unknown prehistoric tsunami had been identified using a wonderful mix of evidence, key to which was LiDAR and what was up until that point a theoretical tsunami geomorphology assemblage. However, to be honest it seems a little like cheating since there was a suggestion that something might have been there—Rakitauneke and his tidal wave were asking to be investigated.

In this case, tsunami geomorphology was found first, albeit with hints provided by poorly documented Māori oral records and some archaeological observations, and the geology only came later. It is an interesting case in many ways, a case that was mirrored by similar work carried out to the north where there was a far stronger signal of earthquake activity but no reported evidence of tsunamis.

This story starts on a dark, wet night on the southeast coast of the North Island of New Zealand. A documentary film crew arrived to capture an early morning vista that, assuming the sun was going to shine, would be worthy of the start of their episode about tsunamis. To set the scene we really have to understand all about the tsunami before we can reveal what put us on the scent in the first place.

It is a striking coastline, steep hills backing a narrow coastal plain striated by a series of gravel ridges running parallel to the coast. Successive earthquakes had uplifted the coastline in sharp jolts a few meters at a time over the past six or seven thousand years. Each gravel ridge marked where the old coastline had been before being so rudely lifted up and away to safety. Here we can see evidence for the inexorable uplifting of the land over thousands of years. This was just like being in the army, long periods of boredom with nothing happening punctuated by sudden unrelenting action. In this instance the action seems to occur about once every 400–700 years. The most recent large earthquake (and tsunami by the way) occurred in the 15th century. The tsunami that was generated by this event inundated numerous prehistoric coastal Māori sites in and around the area, causing widespread abandonment of settlements and a movement inland and uphill. At this particular site of interest to the documentary crew, there are prehistoric Māori gardens built on the narrow coastal plain across these uplifted gravel beach ridges and extending a little uphill. The archaeological evidence left behind consists of rows of stones interspersed by "gardens" or empty spaces—in reality while these "gardens" look seductively interesting, in all likelihood it was the stone rows that were the focus of attention for gardening, not the vacant and quite frankly infertile bits in between. The stones raised the crops a little off the ground, away from frost and exposed to more sunlight while, by extending the rows uphill, they were able to access freshwater springs that provided a regular source of water.

These gardens were definitely in use in the 15th century. The stone rows were laboriously built by "borrowing" or mining the gravel from the ridges of the old uplifted coastline, leaving behind what is known in the trade as borrow pits, a fancy name for a large hole in the ground.

This is where the plot thickens. Things are not quite as simple as I have suggested over a large section of these gardens. To start with a large, partly infilled borrow pit can be seen in the middle of the gardens and surrounding it for several hundred meters the stone rows appear to have been partly or totally buried. Near the inland extent of this section of narrow coastal plain up against the base of the hills is a layer of sand up to 70 cm (2.3 ft) thick. There is no need for a drum roll or anything quite as dramatic at this point, but yes, this sand sheet has been identified as a tsunami deposit, a wonderful example of a direct interaction between humans and tsunamis that is over 500 years old. While this is a fascinating example of catastrophic human-environment interaction, identifying the tsunami deposit was not a simple exercise by any means.

And now back to our documentary film crew.

There were three of them, the director, cameraman, and sound engineer. A small group, but that is the way of most documentary film crews these days. This is not the Hollywood end of the business, there are no paid actors, just the scientists doing it because it is part of their drive for public outreach—trying to tell the stories of their science in a way that will educate and inform the public, not bore them to death. Equally, the documentary crew also wants a story that will sell, no one wants to be assigned to the proverbial cutting room floor. Having said that, sometimes it just doesn't work. In a later part of this episode about tsunamis I was being filmed at another location and, after an exhausting day, the presenter and I had set up to do a final piece to camera. The scenario was simple, they would say something like "so this is what will happen if another tsunami struck this coast?" My riposte which would bring the curtain down on the episode was to be "it's not if, but when." Ta da . . . .

Fifteen takes later, "it's not if, but when" was starting to sound like a cry of desperation from a dying man. This was the last day of more than a week of filming, we had an excellent repartee going, we were all comfortable with the camera and multiple takes to "get it right," but this one simply refused to work. The episode ended with the presenter sagely stating "and so this is what will happen when another tsunami strikes this coast." Sigh.

Back to the film crew.

It was a dark and stormy night. The film crew and the scientists were all housed in the rudimentary and only accommodation of the neighborhood, the sheep shearers' quarter. This site was part of a working farm that had recently converted the fields covering the prehistoric Māori gardens into deer paddocks with fences about a mile high and padlocks on all the gates. It was an isolated farm, a prime target for thieves—of livestock, farm equipment, dogs, whatever. It is a sad indictment on society that it has come to this, but that is part and parcel of country life these days.

The drive to the farm is long, sealed roads run out a good 80 km (50 miles) from the property, and the rain just made it even more exhausting. The crew were tired but since it was the first day of filming, they were also keen to enjoy the company, the warmth of the wood burning stove, a decent meal, and a glass or three of red wine. It is worth remembering that this was not Hollywood, no haute cuisine, and certainly no fancy wines. Each person opened their luggage and took out whatever food they had brought along for this one night's stay. Perhaps more worryingly though, everyone had brought along some wine to share. Now I can hear the cogs turning . . . that means one bottle each for the night, doesn't it? Nice try! Due to a slight breakdown in communication each person had brought along a three-liter carton of red wine thinking that they were bringing it for everyone! I'm not sure what this could be classified as—a confluence of cheapskates, a winos convention?

The dynamics for the evening developed organically along predicable lines. The film crew were not going to be outdone by the scientists, who quite frankly just wanted an early night but felt that they had to at least try to do justice to the well-known mantra of geology fieldwork—"work hard, play hard."

It was a late night, or more precisely an early morning. Sundry stumbling and bumbling "good nights" started at about 1:30 in the morning and with an almost audible sigh of relief finished as the last person crawled into bed at about 1:35.

Three hours later, the bare light bulb in the bunkroom was on and five half-humans were gingerly groping around for clothes, grabbing instant coffees, and struggling through the darkness to the SUV that would take us to the site. The sun was due to rise early, around 6:00 a.m., and we needed to be at the top of the hill to catch the glory of the sunrise as its low-angle light displayed the vista below.

To be honest, not a lot of thought had gone into the ascent, it had just been assumed that it would be a steady, slow climb to the top. The one minor factor we had not taken into account, though, was the deer fencing. At about 2m (7 ft) high it presented a nearly impossible task, akin to climbing Mt Everest in bare feet. The gates were locked, the camera equipment was bulky, and the bodies were anything but willing. I will not dwell on the pain and suffering endured in scaling not one but two deer fences—there was no way around them and even if there was we couldn't see it in brutal darkness of the countryside at night. By 5:30 a.m. we were finally at the base of the hill, and there was nothing for it but an inelegant crawl up the ever steepening slope to the rocks above. At 5:55 a.m., we were exhausted and bedraggled, but successfully perched above the vista below, camera set up, and in position for the piece to camera—ugh, how did it come to this? We were knackered and a lightening but cloudy sky seemed likely to spell doom for our early morning exercise, Nature's final laugh.

Miracle upon miracle, the clouds parted, and the sun began to peer over the horizon. Almost there.

Cue the cameraman.

"Oh s**t, I left my fresh tape in the car, I've only got enough for one take."

We peered forlornly through the dawn light to the vague outline of the vehicle an impossible journey away from our lofty perch. Sod it.

And then the sun shook itself free of the earth, the clouds parted, and a glorious view of the prehistoric Māori gardens below came into sharp relief. Every stone row stood proudly out of the ground in the low angled light, and the sand layer that lay over it could be seen draped over the landscape below, bulking up as it piled up against the edge of the hill below. This was not the tsunami geomorphology of an open plain of undulating dunes ripped apart by the waves' progress inland but rather the short sharp shock of material picked up from the coast and spread over the land, filling in the gaps between some of the stone rows, their physical expression buried beneath the sand, the borrow pit partly filled in, the ebb and flow of the water clearly marked by the shape of the deposit below (Plate 4.1).

"Wow," went the film crew.

"Action," went the director.

And one minute later, the monologue was complete and the one and only take had been perfectly executed.

"Cut."

It was done, and we all sat there is companionable silence marveling at the site below as the sun slowly lifted into the sky, flattening the scene so the stone rows gradually disappearing into the vast expanse of the coastal plain below.

This was and is, as far as I am aware, the only time such a scene has ever been captured on film.

Loins girded we prepared for the precipitous trek back down the hill and to summit the deer fences one more time, but since we were so near the top of the hill we decided to have a quick look at the view, just how spectacular could it be? As we reached the top the view both inland and out to sea was everything we could have asked for, it washed away the last vestiges of hangover and it also gave us a salutary lesson in forward planning. A farm track that we had passed on the way out to the site ended a matter of meters away from the summit. The alternative to the entire fiasco of fences, hill climb, swearing, and energy-sapping man-hauling of equipment was but meters away from us, and we had been blissfully unaware.

Next time, if it ever happened, we would be wiser.

All was not finished though because it was obvious that the deposit was there, we had seen it from on high, so we had to take the opportunity to look at it.

No problem, every farm has some form of mechanical digger and in due course it swung into action, carefully avoiding any archaeology. Trench 1, sand, trench 2, sand, trench 3, sand . . . trench 12, sand. Where was the bloody tsunami deposit? Three hours of digging and it was just sand, sand, and more sand. Not the duney stuff, but a very coarse sand verging on gravel, the entire coastal plain was made of it except for the stone rows.

We two geologists went back to trench 1, eased ourselves down onto the side of the trench and sat there looking at it, willing it to tell us its story. We sat there in companionable silence, calm, relaxed, and staring at the opposite side of the trench. It was a long wait, and you could almost hear the cogs in our brains turning, an invisible checklist being sorted through as the eyes processed the scene again and again and again. And then suddenly there it was, subtle, easily missed, but undeniably a different layer of sediment was sitting on top of another. This was a classic case of sand on sand. The only difference, and it was so subtle that it was really only visible when the sun shone on the trench face, was that the deposit underneath was very slightly browner, discolored by a small amount of fine silt—a sort of trainee soil.

While later on we would prove this by analyzing the samples we took, it was a real eye-opener to know that if you spent enough time up close and personal with the ground you could see it. We were learning and we had with quite some difficulty managed to create for ourselves a new pair of glasses . . . again!

5

Strand 4

Chatham Islands—Rekohu/Wharekauri—How Big?

*the first great wave rushed in with such force and terrific noise that the
very foundation of the deep seemed broken up. In ten minutes more,
another wave, more terrible than the former, commenced its work of
destruction and after a like interval, the third and last completed the
catastrophe.*

*Indeed, the full wrath of the ocean seemed to battle with the island
in fierce resolve to submerge it . . . The third wave, which came rolling
in with most awful grandeur and thousand-fold power, bearing down
outbuildings and stout old akeakes [Olearia traversiorum], which
broke and cracked beneath its fury like matchwood, carrying away
young cattle, and scattering the debris of the ruins far away.*

Hawkes Bay Herald, Chatham Islands

The AD 1868 tsunami was a big one for the South Pacific. We have come across
it before in this story and know that while it may not have been generated
by the largest earthquake in the region (that honor is reserved for the 1960
Chilean event), it was particularly bad for some parts of New Zealand—and
that is important for the story.

It is now that we start edging, albeit almost imperceptibly, toward our goal.

The AD 1868 Arica tsunami is instructive because it shows us that when
waves are generated a long way away across the oceans, the size of the tsunami on your front doorstep largely comes down to the nature and characteristics of the earthquake that generated it. We now know that for any large
earthquake generated in the armpit of South America—this is an easy to recognize geographical feature that is actually known as the Arica Bend or Arica
Elbow, which is not a slur on the lovely city of Arica on the very northern

In Search of Ancient Tsunamis. James Goff, Oxford University Press. © Oxford University Press 2023.
DOI: 10.1093/oso/9780197675984.003.0005

boundary of Chile—the resultant tsunami will impact the east coast of New Zealand (see Chapter 1: Figure 1.2).

Arica is only 18 km (11 miles) south of the border with Peru. Given that it sits toward the northern end of the Atacama Desert, the driest desert in the world, it has a surprisingly mild, temperate climate albeit with some of the lowest annual rainfall rates anywhere in the world. It benefits from the fact that it sits at the convergence of two rivers, the Azapa and Lluta, something of a rarity in this desert landscape. As such, it is a city of extremes, with lush valleys providing fruit for export and desert conditions suitable for Mars expedition simulations. It is generally known as the "city of the eternal spring" and is a haven for beachgoers and surfers alike. Indeed, South Pacific storms produce some of the most famous surfing waves in South America, such as the El Gringo or El Brazo. Perhaps more significantly from the point of view of tsunamis, it is an incredibly important port for much of inland South America and in particular for Bolivia. It is therefore extremely unfortunate that such a glorious location is also exposed to large tsunamis generated by giant earthquakes that occur just offshore where two tectonic plates converge—the Nazca Plate plunging down (subducting) beneath the South American Plate.

At around 5:05 p.m. local time on August 13, AD 1868, a Magnitude 8.5 (possibly as large as 9.0) earthquake generated a tsunami with a run-up height of around 15–20 m (50–65 ft).

At this point I feel that it is important to sort out the term run-up height. This does not mean that a massive 15–20 m wave struck the coast but rather that once the tsunami came onshore, whatever height it was, the waves ran up inland to a maximum height above sea level of 15–20 m. The waves are usually not that high when they come onshore but have so much energy that can easily "run up" that high. To put than in context, each story of a building is around about 3 m (10 ft) high, so we are talking the height of a 5–7 story building.

Any of the city that had not been reduced to rubble by the earthquake was destroyed by the tsunami, with the destruction effectively divided between the upper and lower parts of the city. In the upper part of the city, the earthquake had ensured that there was not a single house or wall left standing, the entire area was in ruins, while closer to the coast, and within the run-up zone, the city had been swept clean right down to and including the foundations. The city had been wiped off the map in a matter of minutes (Figure 5.1).

Figure 5.1 Arica after the earthquake (1868)—looking southwest. (Public Domain image: https://en.wikipedia.org/wiki/1868_Arica_earthquake#/media/File:Arica_after_the_earthquake(1868). Repository: Library of Congress Prints and Photographs Division Washington, DC, USA

The fate of *USS Wateree* was discussed in Chapter 4, and it was this event that was responsible for the stranding of this iconic vessel. To this day, the rusting steel boilers of the *Wateree* sit in the sand hundreds of meters from the shore, ironically partially moved back toward the sea by a later tsunami in AD 1877, and finally by humans who placed the remains adjacent to the main highway along the shore north of the city, and closer to the sea, as a National Monument of Chile and a tourist attraction. The only other remains were recently discovered in a hole in the ground at the site where she was moved to by the AD 1877 tsunami—a small piece of rusted metal. And that is all that is left of the famous ship.

There is even less left behind of the US Navy store ship *Fredonia*, a few pieces of rusted bow poke out of the sand on a now-abandoned plot of land previously earmarked for development but now stalled because of this historically significant find. When you think about it though, this is a wonderful example of human folly and our ability to not only forget but to ignore the

devastation caused by past tsunamis. The AD 1868 tsunami is an historical, and well-documented, event. The *Fredonia* was one of many vessels
destroyed that day by a tsunami that picked them up like matchsticks and left
them broken and marooned high and dry inland. Only two of the crew of the
Fredonia survived, and yet we now want to build on that same land that was
ravaged by the tsunami—we never learn.

One story about the *Fredonia* provides an interesting aside, albeit completely irrelevant to our ongoing journey. It was carrying a lot of supplies
when it was wrecked—around US$2 million worth of supplies to support
the South Pacific Squadron—and although the *Fredonia* had been destroyed
many of her stores fared rather better and were strewn along the beach.
Among the supplies was a rather large store of liquor, and it was reported
that "for three days [after the disaster] even the most humble 'cholo' [Indian]
would drink nothing but champagne." A small recompense for losing everything perhaps? In the desperate search for a roof over their heads, some
refugees could be found living in tents made up of whatever material could
be scavenged from the beach with several constructed entirely of maps of
Bolivia—more bounty from the *Fredonia*!

Further afield, other Chilean towns were also badly affected by the tsunami. Over 400 km (250 miles) to the south was the port of Cobija. Before
Arica became Bolivia's access to the sea, this was their only port. It was
destroyed by the AD 1868 tsunami, rebuilt, and then destroyed again by a
tsunami in AD 1877. Not to be beaten, it was rebuilt again because of its importance as a mining port, only to be lost to Chile in the War of the Pacific
in AD 1884 and eventually by treaty in 1904. It is now a sleepy fishing village.
The skeletal remains of this once great port slowly crumble into dust on the
hill above it. In many ways this was probably a better ending than for it to be
destroyed again by the next tsunami.

Our story though takes us even further away, not 400 km (250 miles) to
the south but almost 10,000 km (6200 miles) west, to a little known but fascinating outpost of New Zealand known as the Chatham Islands.

The Chatham Islands archipelago is about 800 km (500 miles) east of
the South Island of New Zealand, its closest neighbor. The largest island,
Chatham Island, covers an area of 920 square kilometers (355 square miles)
and was named after the survey ship *HMS Chatham*, the first European ship
to locate the island in AD 1791. The archipelago is called *Rēkohu* ("Misty
Sun") in the indigenous Moriori language and *Wharekauri* in Māori.

These multiple names merely hint at the turbulent past of the archipelago.
The indigenous Moriori arrived at this previously unoccupied archipelago

around AD 1500 from the mainland of New Zealand and developed a peaceful way of life in contrast to the warring tribes they left behind. However, in AD 1835 members of two Māori iwi (tribes) invaded from the mainland (having taken passage on a British ship) killing, enslaving, and nearly exterminating the Moriori. Tame Horomona Rehe (Tommy Solomon), the last Moriori of unmixed ancestry, died in 1933.

Given their somewhat isolated location in the southwest Pacific Ocean they are, not surprisingly, exposed to tsunami sources on all sides and several historical tsunamis have caused significant damage and death. Tsunamis generated by earthquakes in the armpit of South America are by far the most important sources of these damaging events, although as recently as 2018 researchers had little idea of just how bad these could be. Up to that point the oldest known event was the AD 1604 tsunami, from Arica, that had been identified in the geological record.

Hindsight is wonderful thing, but I must admit to being one of those who failed to recognize the magnitude of any earlier tsunami inundation. This is largely because so much focus had been placed on understanding what actually happened in AD 1868. For that tsunami I found geological evidence in the northwest of Chatham Island at a place called Tupuangi (Figure 5.2). This

Figure 5.2 Chatham Island, New Zealand—Boulder in photo (c) sits just out of the sea and may have been put there by a storm or tsunami (colleague for scale). The white arrows in photo (d) indicate some of the hundreds of large boulders scattered across Okawa Point by the event that inundated it about 3800 years ago (white oval highlights an SUV for scale). (Image: J. Goff)

was hardly surprising since the site we chose to investigate was immediately next door to the ruins of a house destroyed by the event! What was surprising to me, though, was that no one had ever bothered to even look before we did the work in the mid noughties. At the same time, we also found deposits related to the previous event that came from the armpit of South America in AD 1604—an historical event in South America, but prehistoric in New Zealand.

The AD 1868 tsunami struck Chatham Island around 1:00 a.m. on August 15, having taken some 15 hours to cross the Pacific from Chile. It arrived about an hour before high tide, and a map produced at the time by Thomas Ritchie, a landowner on the island, showed that the worst affected areas were along the eastern and northern coasts. People awoke to a loud roar and water surging through their homes as the first wave struck, followed 10 minutes later by a larger, more destructive wave, and then 3–5 minutes later by a third large wave that completed the destruction.

The Hawke's Bay Herald report on the catastrophe is quoted at the beginning of this chapter. Waves up to 6.0 m (20 ft) high inundated up to 6.0 km inland (3.5 miles).

The entire Māori settlement at Tupuangi, where around 70 people lived, was destroyed. Whare [houses] were demolished, coastal vegetation eroded clean away, with only sand, boulders, and seaweed left behind. Three whanau (families) were washed away with their whare and drowned. On the east coast, the settlement of Owenga suffered a similar fate, and sand dunes along the northern and eastern sides of the Island were badly affected.

The number of deaths caused by this event is somewhat uncertain. Interestingly, newspaper accounts at the time are misleading and are divided almost equally between a single death and more than one although the official historical tsunami database currently states that one person died. However, recent research in conjunction with the local community has revealed evidence to suggest that 23–32 people died, a figure which represents the three large families washed away by the tsunami. This markedly different death toll to the official single fatality may have a lot to do with a lack of trust between local Māori/Moriori and the more recently arrived European colonizers coupled with a somewhat racist attitude toward Māori/Moriori at the time. The fact that even a single death was recorded was most likely because it involved a Māori man trying to save a boat belonging to a European (Pākehā).

There was a lot of destruction of the coastal landforms in sand dune areas beyond the settlements. It was here in the northeast corner of the island at a place called Okawa Point that a group of us decided to look for evidence of the AD 1868 tsunami inland from an extensive line of large sand dunes (Figure 5.2). Based upon our marvelous new knowledge—and the new glasses we were obviously now wearing—we should be able to find evidence for this catastrophic event there too.

There are few places to stay on Chatham Island, and the best spot is in the capital, Waitangi, on the western side of the island. This is by far the largest settlement in the archipelago, with a population of about 250, about 40% of the population. For the unwary traveler it is worth noting that almost everything is flown or shipped in to the islands, and as such supplies can be both unpredictable and expensive. A good rule of thumb is to think of a price and double it, although for some food and drink items this is a significant underestimate.

It is only about 55 km to Okawa Point from Waitangi, but getting there means driving for about two hours along tortuous gravel roads, and so carrying out research there comes with a limited window of daylight hours. Driving the gravel roads at night after a long day or a short night's sleep is challenging, made more so by the idiosyncrasies of any rental vehicle that might be available.

During one visit to the island a group of four of us were exploring the northern coastline about an hour's drive out from Waitangi. As we were slowing down to turn into the gateway of a field the car suddenly slumped backward to the left. Imagining a flat tire, and praying that there was a spare, we leaped out to be confronted by a slightly more worrying scenario. The back left wheel was still on, just. It was leaning at a precarious angle devoid of nuts, held on merely by a couple of bolt threads. Each nut had slowly unscrewed itself over the course of the day and then, one by one, dropped off onto the gravel road to be hidden forever. Well, we actually found one nut, it was a mere meter or two from where we came to a grinding, literally, halt. This presented us with an interesting conundrum, one nut, one wheel (oh, no spare tire by the way, but that wasn't really the problem just then), one hour of gravel road driving to get back to base camp. We cannibalized, or borrowed, one nut from the each of the three other tires, which then left us with four nuts on each tire, and we limped carefully back to town and the hotel (restaurant/car rental/shop/etc.). We arrived slightly weary, slightly shocked from the experience—what if we had actually been driving fast? We were also quite rightly seeking some sense of apology for the poor state of the rental. After regaling them with our tale of woe and our near-death experience, the

lack of a spare and the tale of the nuts, there was a sense of having unbur-
dened ourselves, and we awaited affirmation of our scary experience laced
with profuse apologies.

"You didn't find the other nuts then?" was the sole reply.

In hindsight we should have expected this, after all, nuts are a precious
commodity on the islands, tires are too, so why bother with a spare, and
incidents such as this were two-a-penny, so you just harden up and sort it
out. With a mix of chagrin and confusion we ambled down to the bar and
had a drink to wash away the rigors of the day—there were more important
things to worry about, such as whose round was it? Life is too short to worry
about this other stuff when you are on the islands.

Okawa Point is a gorgeous spit of land on the eastern side of Chatham
Island that sticks out into the Pacific Ocean and says, "hit me with whatever
you've got, I can take it." Tucked into the leeward, southwestern side of the
spit is the start of a long coastal dune system that forms the entire coastline
of Hanson Bay, which stretches 35 km south to the town of Owenga and beyond. This forms the majority of the east coast of Chatham Island. It is a wonderfully isolated and gorgeous stretch of coastline.

The dunes at the Okawa Point end of this system are high—up to 12 m
(39 ft)—and are about four layers deep, but all piled up into each other, so
it is hard to tell exactly. A small intermittent stream that meanders along
an almost gorge-like channel through the dunes from the wetland behind
emerges at the coast a couple of hundred meters southwest from where the
dunes start. We figured that if the AD 1868 tsunami had struck this coast,
then the only place we were going to find something would be close to where
the stream enters the dunes from the wetland behind them. After all, there
were no obvious "blowouts" or tsunami geomorphology, so the dunes were
not breached and must have done a good job at stopping any tsunami waves.
To the northeast of the dunes on Okawa Point there was simply a fairly desolate landscape with not much to talk about. It was almost flat, with a coastal
gravel ridge instead of dunes, not many nice places where we could dig holes
to look for deposits, and an assortment of rocks and boulders littering the
area. We figured that the wetland was the place to look.

The weather was hot. Sheltered behind the dunes, we were out of the ever-
present wind, though we were soon in a battle to avoid being shriveled into
prunes by the baking sun. In anticipation of this great expedition we had
shipped across our coring equipment hoping to core down to great depths
into this ancient landscape. This coring equipment was of a pragmatic design,

honed down to an almost brutal simplicity so that it could survive the vagaries of geological fieldwork. It was a vibracoring system, the basic premise being that you get a bit of aluminum pipe, attach some type of vibrating system to it, hold the pipe up vertically, turn on the vibrating system, and vibrate the tube into the ground inch by inch. The pipe essentially shakes its way down— if the ground is slightly wet and the sediment is nice and fine like mud then the process proceeds like a hot knife through butter. If the ground is dry and sandy, then it is a fight for every inch of progress.

There are an almost endless variety of coring devices that can be used, but this one seems to be the best one to cope with almost any type of sediment, although it has its preferences. In a bit more detail, the lightweight aluminum tube is usually a 6–7 m (20–23 ft) length of irrigation pipe clamped onto the vibrating system, which is a strange snake-like object with a vibrating head on it that is usually used to vibrate the air bubbles out of cement and concrete. Once assembled, it is just a case of providing the motor to drive it and manhandling the pipe into a vertical position above the point you wish to core.

The whole system can be operated by just two people, although three or four is better. In a two-person operation, one person stands by the motor varying the power as the progress of the pipe into the ground varies with the type of sediment it encounters, more oomph to get it through the sticky bits and easing off the throttle if it goes down quickly. The other person manhandles the vibrating head to ensure that the pipe goes down vertically and not at an angle, rotating the tube to help it penetrate or cut through the sediment. This can be a slightly challenging process if progress is slow, necessitating the application of downward pressure on the vibrating head (that is clamped to the pipe) either by pushing it down (do this for many years and you run the risk of getting vibration white finger or Raynaud phenomenon, which affects the blood vessels, nerves, muscles, joints, and connective tissue of the hand, wrist, and arm. Typically, when exposed to the cold one or more fingers will turn white and numb and, on rewarming, blue due to a sluggish blood flow), or if you like by sitting on it. On one occasion this led a female colleague of mine to volunteer enthusiastically considering it to be the largest vibrator in the world!

Back to coring.

The amazing thing with Chatham Island is that there is so little sediment. So why on earth we decided to core I have no idea. Ignorance is bliss, I guess. The rule of thumb of around 1000 years per meter of sediment can be thrown out of the window here. The key issue is that there are no rivers on Chatham Island and so there is no real sediment source other than what the sea can conjure up from time to time.

The sediment does offer some help though. There is one very useful layer of sediment here that helps to date things in the Chatham Islands, the Kawakawa tephra. This is typically found anywhere between about 50 cm (1.5 ft) and 3 m (10 ft) underground. The Kawakawa tephra or ash layer was deposited in the aftermath of the Oruanui eruption of the Taupo Volcano in the central North Island of New Zealand some 1000 km (620 miles) to the northwest. This was the world's most recent super eruption, one with a Volcanic Explosivity Index (VEI) of 8, the largest recorded value on this index. To qualify for a VEI of 8, the volume of deposits for that eruption must have been greater than 1000 cubic km (240 cubic miles), in fact it totaled somewhere around 1170 cubic km (280 cu mi). To put those kind of numbers in more of a human related perspective, the estimated volume of the 828 m (2717 ft) tall Burj Khalifa in Dubai, the world's tallest building, is around 57,522,400 cubic feet which converts to about 0.0004 cubic miles. So the amount of material that erupted out of the Taupo Volcano was equivalent to about 700,000 Burj Khalifas—not bad.

One of the results of this eruption was the formation of Lake Taupo (616 km^2, 238 sq mi in area and 186 m, 610 ft deep). This partly fills the large caldera generated by the eruption. It is now a tourist attraction, where you can partake in a seemingly endless number of recreational activities, but just remember—you are in lake that sits on top of a volcano. That might add a little spice to your early morning swim.

Oh yes, I forgot, the eruption happened about 26,500 years ago. Which, in the context of our coring means that if we cored in the wrong place, one meter of sediment would equate to about 50,000+ years! Not much help if we are trying to find evidence for the very recent AD 1868 tsunami. Fortunately, it was not quite that bad, but even so there was not a lot of sediment. Our longest core only managed to creep just over 3 m (10 ft) down, and the sediment was already 20,000 years old just over a meter down. But we did find something, not a lot, but evidence from the core taken closest to the stream exit showed that a really thin 1 cm (0.5 inch) thick layer of dune and beach sand had been forced inland by the tsunami. Not much excitement really for a week of fieldwork.

BUT perhaps more fascinating and not something we necessarily thought of at the time, the dunes here were not blasted away by the tsunami even though the waves were up to 7 m (23 ft) high. The dunes rise to at least 12 m (39 ft) above sea level and so, as in many other places in the world, they protect the land from being inundated by the sea. Even if the waves had been higher, they still would not have managed to get inland past the densely packed 150 m (500 ft) wide swathe of dunes.

That is all logical and fine, though it raises an interesting question. When we explored the dunes, we found masses of pebbles and even large slabs of rock scattered through them up to 10 m (33 ft) above sea level. Where the hell did they come from?

Ah—now that is indeed an interesting question. They are too high up to be related to the AD 1868 tsunami and they are certainly far too heavy to be windblown. They are composed of rounded and angular material mixed with shells and whale bone fragments that have come from just offshore. Somehow, they also managed to get a good 100 m (330 ft) or more landward into the dune system. This last point is possibly quite useful in helping us come up with a vague idea of when this might have happened. The dunes here are all bunched up together, but if you look elsewhere around Chatham Island you will find that dunes formed during four different phases or time periods. The earliest, and biggest, of these was the Te Onean Depositional Episode between about 5000 and 2200 years ago.

Hmmm—so, if the pebbles, shells, whale bones, and so on are all sitting in these highest Te Onean dunes quite a distance inland then we can suggest that they were probably thrown up there by a very large event, undoubtedly a tsunami, sometime between 5000 and 2000 years ago. If it was earlier, then the dunes would not have been there, and if it was later, we would expect to see them scattered all over the younger dunes as well. OK, so this is not much to go on but at least it tells us that this event, wherever it came from—we will get there in a minute—was not only prehistoric but was also bigger than anything else that has hit the place since, or else they would have been washed away. But if it was that big, however big "that" is, there must be other evidence for it. After all the event in AD 1868, which now seems by comparison moderately small, managed to obliterate an entire settlement and also leave behind decent geological evidence elsewhere.

These ponderings represented the end of that fieldwork and with worldwide events such as the 2004 Indian Ocean tsunami to occupy our time the conundrum of the Chatham Islands was left to fester for a while. Not forgotten, it lingered in the back of the mind, occasionally popping back into the brain cell to remind me that this was an itch that would not go away.

Fast forward to 2011 and the devastating Japan tsunami, the first time such an event was captured live on TV, from a helicopter. The devastation was awe-inspiring and truly brought home to both scientists and the general public alike the need for us to better understand these beasts. We are all still working hard to do so. On one field trip I visited Settai valley on the main Japanese island of Honshu to investigate an unusual tsunami deposit comprised of absolutely everything from the finest mud up to boulders as large as 461 tonnes (453 tons) and almost 13 m (43 ft) long. The tsunami got to at least 28 m (92 ft) above sea level and inundated almost 2 km (1.2 miles) inland. This was huge.

There was a niggling alarm bell ringing in the back of my mind, I had seen something like this before but could not quite put my finger on it. And then came one of those eureka moments. That night, sitting in a small hotel room in Japan staring at my laptop, it all became so blindingly obvious that I am almost too embarrassed to admit it. Fortunately, I still had all my old field photos from endless trips around the Asia-Pacific from about the beginning of the millennium onward, largely because I simply have never quite got around to putting them elsewhere. I know they take up so much space the computer might crash and I would lose them all, etc., etc., etc. but I still have them on my laptop to this day!

As a salutary warning to those like me who just let these things flounder in the "I will do it one day" folder, I should just mention that I have lost an entire fieldwork's worth of photos, not once but twice. To this day I have no idea how they disappeared from my computer. The first set was from Halapē, a gorgeous site in the southeast of the Big Island of Hawai'i, in Hawai'i Volcanoes National Park. If you ever visit you can't miss it, largely because to get there you have to make a conscious choice to hike the Halapē Trail. The approximately 12 km (8 mile) trail that we took is commonly called a grueling, hot hike. The park information warns of intense sunlight, wind, and high temperatures that can lead to dehydration, heat exhaustion, or stroke. The hike should not be attempted in the heat of the day between about 10 a.m. and 2 p.m. Wear sunglasses, sunscreen, a hat, drink plenty of liquids, and there are NO trees to provide any shade (actually there was one that I recall). Oh yes, and the trail is steep and rocky dropping 915 m (3,000 ft) from start to finish (and back up again of course). I have done it, twice so far, and yes, it is hot. We took plenty of liquid, and each of us had a large frozen pack of Marguerita mix (including Tequila) in the top of our backpacks—once a geologist, always a geologist! The logic of this particular plan was cunning. As it slowly thawed, the Marguerita mix kept the back of our necks cool and once we arrived at our destination and set up our tents, it was just the right temperature to drink—perfect. We took plenty of water too, by the way.

Halapē is a gorgeous little beach with a few dead coconut tree stumps. These are the last remains of a gorgeous little grove of coconut palms that used to grow there until they were unceremoniously drowned in the sea when the southern flank of Kilauea, the island's active volcano, slid a little into the sea back in 1975. The tsunami this little slide generated was almost 8 m (26 ft) high. This wave picked up campers and carried them inland toward the base of the cliff where they were stuck until the waves subsided, many trapped in the lava crack along with boulders, vegetation, and so much more. Sadly, one man died from being buried in the sediment, and others suffered numerous injuries, and swallowed seawater and sand while being dragged across the jagged lava. Only when you stand at the beach do you get any sense of how completely exposed you are to any even minor tsunami. There is nowhere to run, nowhere to hide, and I spent most of my time constantly checking the state of the sea while we tried to make scientific sense of what had happened there. It was almost 30 years after the event, but you could still get a sense of the carnage, if only through the lens of the boulders the tsunami left behind. All were measured, studied, and catalogued. All were photographed too, and all those photographs evaporated when my laptop was "upgraded" by an over-zealous computer technician, although apparently they didn't touch a thing! I guess this must be an excuse for me to return before the bones start to creak too much.

The other occasion I lost all my photographs was very soon after a similar tsunami exploration, this time along the wild and very photogenic west coast of Canada. Like Halapē, there are plenty of tales to tell about that trip, but let us just say that we worked very hard and played just as hard too. The stories will doubtless come out in the end, unlike the photos that evaporated—yet again courtesy of a third party.

Meanwhile, back in the small hotel room in Japan, there were the photos from our Okawa Point fieldtrip laid out before me. Yes, when we did that work we had had "new glasses" on but at the time those had been fine-tuned for tsunami geomorphology and coring. So we knew where to look and what to look for, but it had been hard work because the geomorphology wasn't quite right. It had all been squeezed in at the coast and the wetland wasn't exactly yielding a long record of any action, but we did get something out of it and we did have the mysterious pebbles and junk on the high dunes. We felt we had at least achieved something after all that effort.

But now, in that Japanese hotel room, the photos showed it all. There was a lovely one of me posing on the flat lands of Okawa Point to the northeast of the dunes, standing next to a massive boulder that was almost like a 3.5 m cube. It was huge. And a nice photo.

Duh....

What had we said? Something like, "Okawa Point was simply a fairly desolate landscape with not much to talk about. It was almost flat with a coastal gravel ridge instead of dunes, not many nice places where we could dig holes to look for deposits, and an assortment of rocks and boulders littering the area. The wetland was the place to look."

There was an amazing scatter of boulders across the entire area. It looked impressive in the photos, and the dredged up memories seemed to confirm what the photos showed (Figure 5.2). We had been so focused on what was underneath the ground that we hadn't even bothered to really think too much about the huge boulders littering the landscape. But now there was another new pair of glasses sitting firmly in place and the boulder field at Settai looked remarkably like the one at Okawa Point. Having said that, could this simply be a case of equifinality? Storm boulders creating the illusion of a tsunami?

Needless to say, there was only one way to find out, and that needed funding.

It is difficult to describe the frustration that researchers feel when they know that there is some really important work that needs doing and yet funding agencies fail to see it, however sexy you make it sound. An interesting case in point here—apart from Okawa Point, which we will come back to in a minute—is the Tongan archipelago in the southwest Pacific. The islands are wonderful, the people are friendly and keen to learn more about their hazard history, and boy is there some history.

Back in about 2008 I had the pleasure of going there for a conference and met some Americans who were studying a bunch of huge coral limestone boulders near the village of Fahefa on the western side of Tongatapu Island. The largest of these was at least 9 m (29 ft) long and weighed in at about 1600 tonnes (1575 tons) (Figure 5.3). It is a beast! It turns out that this little gaggle of boulders were thrown up 10–20 m (33–66 ft) above sea level by a tsunami. Pretty cool. But while those are somewhat iconic boulders, there are

Figure 5.3 Huge coral limestone boulder near Fahefa, Tongatapu Island, Tonga. (Photo: Fred Taylor; with permission)

others on that island too, on the eastern side. This is important because the eastern side is not a million miles away from what is a known as the Tonga trench, a very active subduction zone (we have talked about it earlier). This is over 1300 km (800 miles) long and over 10 km (6 miles) deep, the deepest trench in the Southern Hemisphere, where the Pacific Plate is subducting beneath a mishmash of bits of pieces associated with the Australian plate. This subduction is happening at a rate of up to 24 cm/year (9.4 in/year), the fastest on Earth. And yes, it generates not only some very large earthquakes but tsunamis as well. There has been an ongoing debate for many years as to exactly how big and how often earthquakes and tsunamis occur along this subduction zone. This debate can get a little heated from time to time with various arguments ranging from, "holy cow if the whole thing goes it will greater than a Magnitude 9" (indeed, I have argued for as many as three such events over the past 3000 years or so) to "it is never going to be that big an event because it is broken up into segments." This is a very important debate because places such as Auckland, the largest city in New Zealand, are a little over an hour by tsunami from the Tonga trench, hardly any time at all to evacuate a population that is largely unaware of the threat. A bit farther to the west is the east coast of Australia and its population, but then let's also not forget Tonga, New Caledonia, the Cook Islands, and so on, which are all much closer to the trench.

It would seem incredibly logical to be very concerned about this uncertainty and do some urgent research to try and get to grips with the problem. Where would be the best place to look? Tonga! As and when a big tsunami does happen, its first port of call will be the Tongan archipelago. What about the unstudied boulders on the east coast of Tongatapu? What about the vast complexes of wetlands and bays that could be cored to look for geological evidence? Again, these have never been studied. What about funding for this urgent research? I have tried to get funding on and off for the past 20 years with no success. Others have also failed.

Serendipity is a wonderful thing sometimes though. The 2022 eruption of the Hunga Tonga–Hunga Ha'apai volcano that generated a Pacific-wide tsunami occurred at just the moment that a funding proposal was submitted. There may finally be some light at the end of the tunnel. This proposal was made along with researchers from Australia and, glory be, it was successful! This information is hot off the press, and so nothing has been done as yet. Strangely though, there is also now some Japanese funding. Both cases almost scream to me saying that these represent the classic "reactive" funding which while wonderful to receive makes you wonder what progress could have been made if there had been "proactive" funding two decades earlier. It may seem strange to obtain funding from Japan because after all, before the 2022 eruption it seemed highly unlikely that Japan would be affected by tsunamis from this neck of the woods. But Japanese researchers are, not surprisingly, very interested in understanding all things tsunami related and as such work in Tonga on all sorts of tsunami sources would build upon research that they have already done in the South Pacific, in particular, at Okawa Point!

Again, it may seem odd that Japanese researchers would be interested in a bunch of boulders stuck down in the back of beyond on Chatham Island, but there are two key points here. First, its resemblance to the boulder deposit at Settai makes it a very interesting topic to study. Could those at Okawa Point have been put there by a similar type of event? If so, where did it come from? Second, the boulders at Settai are by no means the only ones they have in Japan. There are some incredible boulders in the Ryukyu Islands, the southernmost outliers of the Japanese archipelago, almost as far south as Taiwan, which sit adjacent to the Ryukyu trench, another subduction zone associated with the Pacific Ring of Fire. This area features a battle between the Philippine Sea Plate subducting beneath the Eurasian Plate. And yes, again, this generates some pretty big earthquakes and associated tsunamis. The last biggie was the AD 1771 Great Yaeyama Tsunami (明和の大津波, the Great Tsunami of Meiwa). This caused thousands of fatalities on the islands but also transported some massive boulders that weigh up to about 210 tonnes (206 tons) up onto the reef and as much as 1290 m (1400 yards) inland. Interestingly, a separate group of smaller boulders didn't get as far, and it was discovered that these were all put there by typhoons (hurricanes/cyclones). Two useful bits of information come out of this work on the Ryukyu Islands. First, on a good day it is possible to differentiate between boulders tossed inland by tsunamis and those put there by storms. Second, there was yet again a subduction zone involved, similar to the one that caused the 2011 Japan tsunami.

So, quite wisely, the Japanese are eager to learn more about such deposits because it helps to explain just how powerful a tsunami can be when it doesn't have any sand to play with, and that is where Chatham Island comes in. Pretty much all the sand on Chatham Island got there as the sea level rose when the ice sheets melted after the last glaciation. As the sea rose it effectively bulldozed any available sand ahead of it until about 6000 years ago when the sea level got to around about where it still is today, though it is now rising again after recent human tinkering with the climate. Any movement of the sand dunes on Chatham Island has been as result of periodic disturbances. Massive erosion by the sea then moved the sand back offshore into the clutches of the ocean, only for it to be gradually put back onshore in a new phase of coastal dune building. As we mentioned earlier in this chapter, the first big dune building phase was the Te Onean Depositional Episode between about 5000 and 2200 years ago. We once thought that these massive disturbances of the coastal dunes were caused by large storm events, but more recently most have been linked to major tsunamis.

Okawa Point seems to be an interesting, and dare I say it, worthy recipient of some research funding. Are the boulders on Chatham Island tsunami-related, storm-related, or both? When did this event or events happen? How does any of this fit in with what the dunes have been doing? Oh yes, and if there was a tsunami involved where did it come from? You see, the one curiosity about Chatham Island is that unlike the Ryukyu Islands or mainland Japan, the only moderately close subduction zone is the Hikurangi Trough just off the east coast of New Zealand's North Island, where a part of the Pacific Plate is subducting beneath the Australian Plate. While this is quite a long way off, some 600 km (370 miles) to the west of Chatham Island, that is no distance at all from the perspective of a tsunami, BUT the tsunami would arrive on the wrong side of the island. Fancy computer models have shown that the wave would not be a large enough to move boulders once it had found its way around to the relatively sheltered eastern side where Okawa Point is to be found. Curious.

While Tonga has only just registered on the radar of research funding agencies, hopefully the Japanese and Australian funded projects will make some significant progress. So there you have it. It is nigh on impossible to get funding from any local agencies (OK, the Australian money has just come through, but NOT from a traditional funding source) but rather from a country on the other side of the Pacific that can recognize the value of the work (n.b. recent French-funded tsunami research has produced some great results from Tonga—yet another country that sees the value of such work!). A sad state of affairs in some ways, but there are many demands on research funding, and I have only my myopic vision of what is important. However, what a wonderful opportunity it will be to work with our Japanese colleagues again, and away from Japan again.

And so, armed with Japanese funding it was back to Okawa Point.

The logistics of such a trip to a remote outpost in the Pacific can be a bit daunting, but Chatham Islanders are a most welcoming and sincerely interested community. They know they are remote, they know that funding to work there is as rare as hen's teeth, they also know that when it comes around to New Zealand and the tsunami hazard, their track history has been a rough one. In historic times, apart from one locally large tsunami that affected a very small part of the eastern side of the North Island of New Zealand, the Chatham Islands have been the first port of call for the top three tsunamis to have struck the country. All of these came from South America, and they all struck the islands a good hour before mainland New Zealand. In a sense the Chatham Islands are now seen as the canary in the coal mine, the early warning system for everyone on the east coast of mainland New Zealand of the nasty wave that is coming. While there have been many technological advances since the earliest events of the 19th century, it is still often difficult to convince the public that a big tsunami is coming, but any proof of the destructive power of the waves on the Chatham Islands is a powerful motivating tool.

This is not much help for the Chatham Islanders though, and they know it. Amazingly, yet again, no other research has been carried out since our earlier work managed to push the evidence of tsunamis on Chatham Island back to AD 1604, and even that was a pretty preliminary effort, so as usual, so much more could be done. Not to labor the point, but it was somewhat ironic that the team studying Okawa Point was comprised of Australian and Japanese researchers—not a single one from New Zealand. It was only on the second research visit that we managed to get a New Zealander on board, and even then it was something of a self-funded effort. Hey ho.

The landowners could not have been more helpful—"here's the key to the gate, just give it back when you've finished." The island's emergency management representative did all she could do to help as well. The locals were all interested in finding out what we could tell them. All that remained was for the work to commence.

One of the great things about rental cars on Chatham Island (notwithstanding our earlier adventure) is that you can pretty much take them anywhere. This was a major bonus given that the nearest road access was a good 4 km (2.5 miles) away from the site. Having said that, the route from the road to the site was a challenging mix of farm track, fields, sand dunes, and the odd stream. Probably slightly faster by car than on foot, with the most significant advantage being that nothing needed to be carried too far.

Over 270 boulders were measured, and more than 30 trenches were dug to study any fine sediments there might have been associated with them. Boulders up to 98 tonnes (96 tons) in weight were traced up to 800 m (875 yards) inland. Just as in the Ryukyu Islands there were two distinct groups of boulders, a scatter of boulders that seemed to march inland across the undulating surface of the point and another lot that seemed to be loitering around the edge unwilling to fall back into the sea but equally unable to make any more progress inland.

The boulders scattered across the point were associated with a thin gravel layer that extended well over 1 km (0.6 miles) inland, which we dated to around 3800 years old. These boulders and the gravel layer were put there by a tsunami. It was hard to tell about the other boulders closer to the sea, they could conceivably have been put there by one or more really big storms. We noted two other cool things at Okawa Point as well. The first thing was that there were no sand dunes here at all, but on either side of this huge boulder scatter there were dunes, one lot being the ones we had looked at all those years before. These dunes had survived the onslaught, but the wave that brought the boulders inland had swept any dunes out of the way, pushing the sand across the point and into the water beyond. This would later be brought inland by the sea again, in a new phase of dune forming.

The second cool thing was working out where this tsunami came from. To do this we used mathematical models to compute tsunami wave characteristics from different potential sources. It couldn't have been a local source, there wasn't one. It was too flat offshore so it must have come from a more distant location. Models were produced, numbers were crunched, and a worst case scenario tsunami coming from the Hikurangi Trough (just off the east coast of the North Island) was, as I mentioned earlier, found to produce some really nice waves along the western and northern shores of the island, but as we thought, not much at all at Okawa Point. Hmmmmm, where else could we look? The obvious answer was South America, but even the massive AD 1868 tsunami that was up to 7 m (23 ft) high here did not manage to move anything like the huge mass of boulders we found at Okawa Point. We did however manage to find evidence for the AD 1868 tsunami on this exposed, undulating, boulder-strewn surface of Okawa Point—a really thin gravel layer, less than half an inch thick.

Much head scratching followed until we finally had to go for what we considered was probably an unrealistic option, the biggest thing that could "in theory" come out of South America. We knew that the debate for how big an event could be in the Tonga Trench was (and still is) raging and hinges largely on the inability of a fault to rupture across different segments of the fault. So to suggest a similar segment-crossing scenario for a really big one in Chile might be pushing things too far and generate even more debate!

These segments are a bit like stations on a railway line, a train (earthquake/fault rupture) can move between two stations but cannot go through them and make the entire journey. However, to continue the analogy, what we were saying is that from time to time you get a fast train that goes through a station (or more than one), rather than stopping there.

So, to hell with it, who cares? Working backward from the size of wave that would be needed to move the boulders, we determined that a tsunami that would be big enough at Okawa Point would have to have been generated by something like a Magnitude 9.5 earthquake in the armpit of South America—northern Chile. Coincidentally, at almost exactly the same time we floated this as an idea, a group of colleagues in New Zealand had done their own modeling to find out where the nastiest tsunami to hit the

Chatham Islands might come from. They ignored segments on faults as well and, lo and behold, their worst case model suggested that it would be from a Magnitude 9.485 earthquake in northern Chile. How about that. All at once we had the deposit AND we had two different lines of mathematical evidence that pointed to what appeared to be the only source.

After a little bit more hunting around the South Pacific we added more to this story. We found more geological evidence for a large tsunami around 3500–4000 years old in other places around the Pacific, such as mainland New Zealand, Australia, Vanuatu, and Samoa. There was a little icing on the cake of all these findings too. A small earthquake and tsunami associated with the northern Chile region, the Magnitude 8.1 Iquique earthquake occurred on April 1, 2014, and we had detailed tsunami wave amplitude information across the entire South Pacific Ocean to look at. As the tsunami moved out across the Pacific Ocean, five fingers of higher wave heights reached out westward across the ocean heading directly for the Chatham Islands (and New Zealand), Vanuatu, Samoa, the Antarctic Ocean, and the Cook Islands/French Polynesia. We had unknowingly identified three countries (New Zealand, Samoa, Vanuatu) most likely to have been struck by this event by looking further back in time into the geological record.

A few days after our return from the fieldwork at Okawa Point I received that unexpected email from Chile—serendipity in action—"we are currently working on the hypothesis of a mega earthquake and possible tsunami occurring in northern Chile around 4000 years ago." Could I help?

The work that started this book has now come full circle, a tenuous link was reaching out across thousands of kilometers of ocean and between countries. Could this be the missing link? Could New Zealand and a wide array of South Pacific countries truly have experienced a tsunami so bad that it devastated their coastal landscapes? We have seen boulders, massive ones, moved by tsunamis generated by large earthquakes that occurred a few tens of kilometers offshore, but this tsunami came from around 10,000 km (6200 miles) away. This modeled event was a worst case scenario that looks great on paper. But could it actually happen? Did the last one happen 3500–4000 years ago? How often has it happened? So many questions.

Maybe, just maybe, this could be the game changer we were looking for, taking mathematical models out for a spin and showing that it did actually happen. If that was the case, it would have serious repercussions for our understanding of the tsunami hazard throughout the Pacific. If you could get an earthquake that big in northern Chile when many experts said it was impossible, then what about all those other places where it was apparently impossible too, like the Tonga trench?

To be fully armed for the harsh coastal environment of northern Chile, though, it is important to have a handle on how tsunamis fit into the bigger picture of environmental change. After all, if we couldn't find any tsunami deposits there—a distinct possibility in such a brutal environment—then what could there possibly be to fall back on and use as evidence?

6

Strand 5

Just Part of the Problem

To monitor the origin of the most recent beach ridge to arise, high al-
titude imagery of the area was digitized and computer overlaid for
1944, 1955, 1961, 1972, 1982 aerial photographs, 1975 and 1978
LandSat MSS imagery, and 1984 Large Format Camera Space Shuttle
photographs . . . the imagery indicates that a new beach ridge arose
after 1970, and was in place by 1975, ... Formation of the most recent
cobble ridge can be interpreted as the synergistic consequence of: a) the
7.7 seismic event of May, 1970; b) the strong El Niño of 1972–73.

M.E. Moseley et al, 1992.

The 1970 Magnitude 7.7 Ancash earthquake (aka the Great Peruvian earthquake) occurred offshore from the coast of Peru. From our point of view, it
is a neat example of where tsunamis "fit" within the bigger picture of environmental change. In this case, because the earthquake was relatively small
and quite deep (40+ km under the seafloor) it didn't disturb the seafloor that
much and so only generated a small tsunami. But that is not overly important
for us because it is the associated after-effects that we are interested in here,
and it provides a good example of those.

The most significant immediate after-effect was that the groundshaking
caused by the earthquake generated a massive landslide that incorporated large quantities of snow and ice into it, a deadly slurry of material
that engulfed the town of Yungay and caused as many as 70,000 casualties.
It is considered the world's deadliest avalanche. In all, the groundshaking
from the earthquake affected an area of about 83,000 km², triggering a vast
number of landslides, and destroying regional infrastructure and 80–90% of
all the buildings in Chimbote, Carhuaz, and Recuay. It affected around three
million people.

In Search of Ancient Tsunamis. James Goff, Oxford University Press. © Oxford University Press 2023.
DOI: 10.1093/oso/9780197675984.003.0006

Such devastation is a common theme for large earthquakes, and if they occur offshore and are shallow enough to move the seafloor then they also come with an associated tsunami—the bit we are interested in. Such tsunamis and landslides are all part of the immediate catastrophic after-effects most of us have come to expect and understand. The longer-term after-effects can be equally catastrophic and can also be traced back to the earthquake that spawned them.

First of all, in this Peruvian case there was not just the one massive landslide but a seemingly endless number of them up in the Andes. The combined effect of all these landslides was the production of vast amounts of sediment that fell into the upper catchments of the region's rivers. However, because of the desert climate here and the consequent lack of rain the sediment just sat there doing nothing much in particular, simply minding its own business.

That was until the strong El Niño of 1972–1973.

It all started in December 1971 when warm waters from the Gulf of Guayaquil off the northwest coast of South America moved southward along the coast of northern Peru. By February 1972, unusually warm waters were spreading further afield. Near the shore, tropical waters of about 25°C reached as far south as Chimbote and heavy rains started to fall flooding agricultural land and damaging property. The rains continued through April and May and while things improved a little from then on, even a year later sea surface temperatures were 2–3°C higher than average. Only in April 1973 did normal conditions resume, by which time El Niño had done its worst to the offshore fisheries and people's livelihoods. But El Niño's rain had also done its bit.

By March 1972 it was obvious that it was going to be a wet year. Light intermittent rains fell on March 9, 10, and 11, but between March 14 and 17 parts of northern Chile and southern Peru recorded the heaviest rainfall in 47 years. All of the dry or intermittent rivers suddenly filled up, overflowed their banks, and caused massive flooding.

And so here is the scenario that has been played out over millennia. The 1970 earthquake (that generated a tsunami) produced a vast number of landslides. The landslides in turn pumped vast quantities of sediment into the upper catchments of the dry rivers that proliferate the slopes of the western Andes. And while the sediment may sit there benignly through the years, all it needs is a catalyst to get things moving, and when that catalyst is a particularly nasty El Niño then all hell breaks loose. Torrential rain falling on a desert landscape devoid of vegetation is capable of moving huge amounts of the sediment that has been lying around, sending it rapidly down through the river systems to the coast.

In this part of South America the two events are needed though, a large earthquake followed by a decent El Niño—tectonics and climate combining to unleash hell. Once the sediment gets to the sea, it is moved along the coast by longshore drift and is brought up onto the shore to form linear beach ridges parallel to the coast. Decades of high-altitude, time-lapse images show this actually happening following the 1970 earthquake, with unstable sand dunes marching inland, covering productive coastal land, overwhelming settlements, and leading to the abandonment of coastal dwellings, while a new beach ridge of sand dunes forms at the coast. This new beach ridge had formed by 1975, that's just five years from start to finish. It could have taken longer if the timings had been off, but five years is like landscape change on steroids—and it can take place over hundreds of kilometers of coastline.

And there we have it—a powerful example of tectonic and climatic extremes combining to make life a living hell in a desert landscape. It was important to go through this historical example because, completely independently from this excellent research, we came up with the same findings for just the same processes operating in prehistory, but in much more temperate regions, not deserts. It also doesn't need an El Niño to unleash hell, it can happen and has happened in so many other parts of the world because all it needs is a bit of rain preceded by a large earthquake to get the ball rolling, and let us not forget, if that earthquake occurs offshore it can generate big tsunamis as well.

That is the point though, the earthquake "gets the ball rolling." We are very good at worrying about one particularly nasty hazard or possibly, if we push it, two associated ones such as an earthquake and a tsunami, and so there is often a wonderful air of complacency among coastal populations, for example, once they have mastered the art of tsunami awareness. They know their escape routes, they know what the warnings signs are, they can self-evacuate—what can go wrong?

Everything.

First of all, if the earthquake that causes the tsunami is big enough, and invariably it has to be pretty big to generate a decent tsunami, then the tsunami just becomes part of what we call a Seismic Staircase of environmental after-effects, or Seismic Driving (Plate 6.1). The tsunami sits within this staircase as just one of a suite of potential disasters along the lines of what happened in 1970. The difference being that in many parts of the world the absence of an El Niño to drive things through really fast means that the timelines for these after-effects can be drawn out over decades and even centuries. These become slow-moving train wrecks that because of their long-drawn-out gestation period tend to obscure the original driving mechanism that spawned them, making people respond to sudden mounds of sand starting to build up on their doorsteps as if they were entirely unexpected, abnormal, and inexplicable. But it is clear to see the cause if you just step back and join up the dots.

The southwest coast of the South Island of New Zealand is a glorious part of the world, a true temperate rainforest along the lines of that found around Long Beach on the coast of western Vancouver Island or the Olympic Peninsula of Washington State. The rainforest is there because . . . it rains . . . a lot . . . as we have seen earlier on in this book.

This is not the case of an El Niño suddenly hammering a land devoid of rain. This is something akin to a constant supply of water, sometimes heavy, sometimes light. In other words, the Seismic Driving system in New Zealand has a permanent lubricant. Not only that, it also has the Alpine Fault system.

The Alpine Fault system is an almost perfect example of how the entire process works when El Niño is not around. And it shows where tsunamis fit into the picture. And by understanding the processes involved we can join up the dots of a series of apparently unrelated environmental changes and human responses. Prehistoric Seismic Driving here goes something like this.

A magnitude 8.0 or larger Alpine Fault earthquake occurs (this happens about every 300 years or so depending upon where you are on the 450+ km long fault). The groundshaking from the earthquake causes a vast number of landslides up in the Southern Alps—and that in turn massively disturbs the forests, knocking over huge areas of trees that had been happily growing on the mountain sides. Either because of the fault movement itself offshore to the south of Milford Sound or because of subsidence/uplift on land to the north, a tsunami is generated that inundates the coast. For those fairly recent earthquakes and associated tsunamis over the past 700 years or so, these disturbances have also caused temporary or permanent abandonment of prehistoric coastal Māori settlements.

These are the IMMEDIATE effects.

Since there is no significant El Niño deluge from time to time the lubricant needs to be readily available to move the sediment that the landslides have dumped into the upper reaches of the rivers that drain the Southern

Alps. Fortunately (or unfortunately depending upon how you look at it), the Southern Alps have the highest annual rainfall anywhere in New Zealand. You can get over 11 m of rainfall a year up in the mountains! To put this in context, it is about the fourth wettest place in the world, with Māwsynrām in India being first with 11.8 m a year, which is not much more to be honest. There they use grass to soundproof their huts from the deafening rain during the rainy season, while in New Zealand they almost celebrate the rain by having tin roofs that amplify the noise!

While there is no El Niño there are two key contributing factors for this excessive amount of rain. First, there is the orographic (mountain) effect. The weather in New Zealand comes from the west across the Tasman Sea, where warm air can pick up vast quantities of water vapor from the ocean. As these air masses then move across the coast they are forced to rise rapidly over the mountainous topography of the Southern Alps. As the air rises over the mountains, it cools. Water vapor condenses and rain falls. As a result, the rain is concentrated on the mountain sides and also increases with elevation, just where the headwaters of all of the rivers can be found. Second, a phenomenon known as Atmospheric Rivers significantly augments these large rainfall totals. These are essentially filaments or long stringy regions of strong horizontal water vapor transport in the atmosphere. They focus even more water vapor than usual onto the mountains and help to boost those rainfall levels even higher.

As the poem quoted in Strand 1 goes to show, West Coasters are pretty laid back about their rainy climate.

While in Peru the sediment had to sit back and wait a while until the massive deluge arrived, a moderate deluge happens almost every day on the West Coast, and if one doesn't then just sit back and wait for a few hours and then there will be one! The second phase of the sediment conveyor belt can therefore begin. Obviously it doesn't all happen at once, and also the landslides will keep going for a while too as aftershocks continue to shake the mountains, but the vast amounts of material dumped into the rivers provides a huge extra slug of sediment for them to move. Imagine if you were water blasting soil and gunk off your driveway and a contractor came along and accidentally dumped a huge pile of extra soil on the driveway. You would eventually get rid of it all with your hose, but it would take a bit of time . . . and make quite a mess.

The rivers of the West Coast carry lots of sediment at the best of time. Indeed, the Haast River that flows out of the Southern Alps transports nearly

20 million tons of it every year, one of the largest amounts in the world, and that is just one of the rivers. Multiply that by say 100 times and you get a sense of how much is coming down the rivers after a big earthquake. This sediment is not all the same size, it ranges from really fine muds to huge blocks of rock, and so they don't all move at the speed. The fine stuff, sand and mud, gets transported pretty quickly to the coast. Since the Southern Alps are close to the West Coast this material doesn't have too far to travel, a few kilometers at best, which means that a massive pulse of fine sediment reaches the coast within a matter of 10–20 years or so.

As we know, coastal processes start to operate then, and the sediment gets moved along by longshore drift, which in this case is to the north. The Haast dune system just to the north of where the Haast River enters the sea is big. It consists of a seemingly endless series of dune ridges parallel to the coast that extends about 10 km (6 miles) along the shore and up to 5 km (3 miles) inland. So it is easy to imagine that anyone who chose to live in or around the dunes would fairly quickly recognize that it was not a particularly great place to be because sand just keeps on being blown through gardens, houses, and anything else that happens to be useful to humans. So if settlements had not been immediately abandoned following the earthquake and tsunami, they would doubtless be once the new slug of sediment arrived at the coast and sand started blowing inland. A delayed effect and again, part of the Seismic Driving process.

Let's leave the dunes there for a second and have a look at the slower stuff, the bigger bits of gravel that take longer to work their way through the system. This material is pretty easily transported by the rivers down the steeper slopes of their course, but as that slope gets flatter the transport of this material starts to enter the phase of being "in the too hard basket," and it gets dumped. This dumping process sees vast expanses of gravel deposited in the river channels, forming thick sheets of gravel or aggradation surfaces. They cover everything—forests die because they are covered in gravel, soils are buried, and any hapless settlement that happens to be in the way is also overwhelmed by this influx. I remember talking about all of this once to a river engineer who worked for a local council in the area. They were confident enough that the stopbanks or river containment structures they had in place would be able to cope with this BUT, and this is a big but, all of their calculations were based on the worst case historical examples, a mere fraction of what will be coming down the rivers after a really large earthquake— bigger than any in historic time.

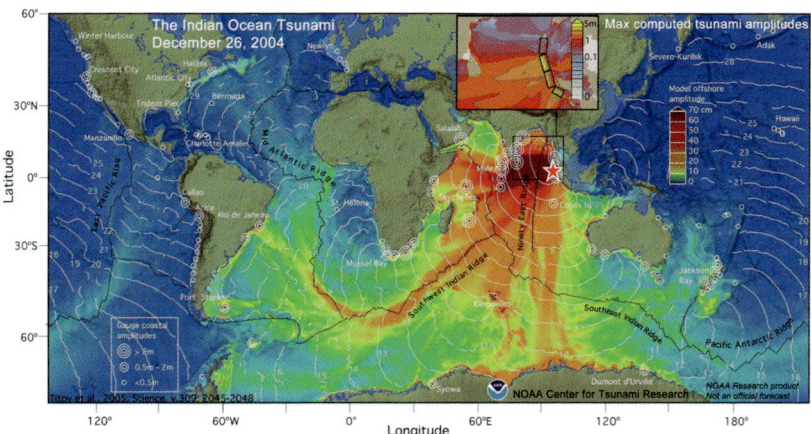

Plate 2.1 Maximum computed tsunami amplitudes around the globe for the 2004 Indian Ocean Tsunami. Note that the tsunami was recorded as far away as Great Britain and the Pacific Ocean. (https://nctr.pmel.noaa.gov/indo_1 204.html. Courtesy of NOAA/PMEL/Center for Tsunami Research. Accessed 27 July 2021)

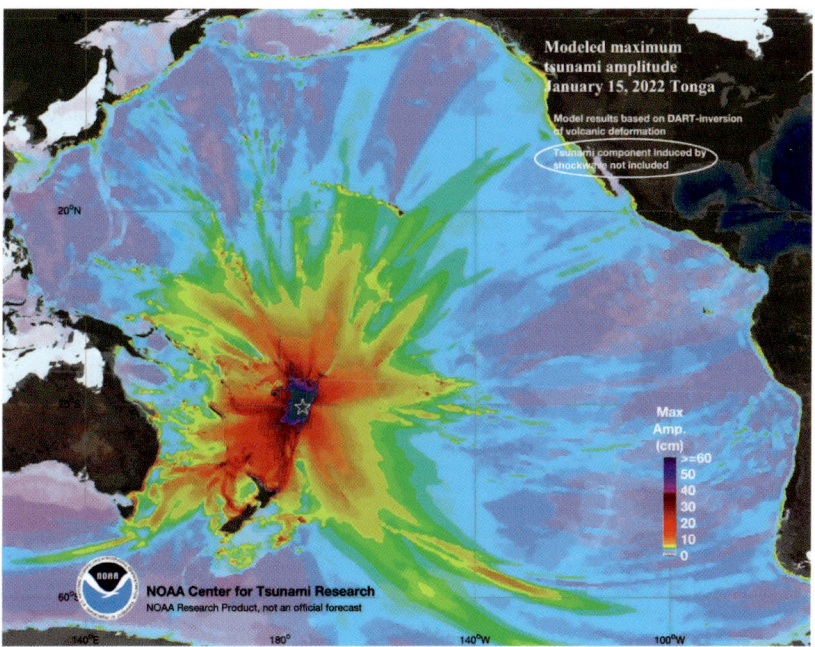

Plate 3.1 Amplitude model for the 2022 Hunga Tonga-Hunga Haʻapai tsunami. (Note: this does not include the tsunami component generated by the shockwave—one of the interesting wrinkles that scientists found enhanced the Pacific-wide nature of this event) (Image courtesy of NOAA/PMEL/Center for Tsunami Research: https://nctr.pmel.noaa.gov/tonga20220115/images/tonga_ maxamp_newinv.png)

Plate 4.1 SE coast of New Zealand showing a tsunami deposit in a low angle light. Arrows show direction of tsunami flow. Dotted line shows the inland extent of the deposit, that partially infills stone rows and builds up against the cliffs. (Photo: B. McFadgen).

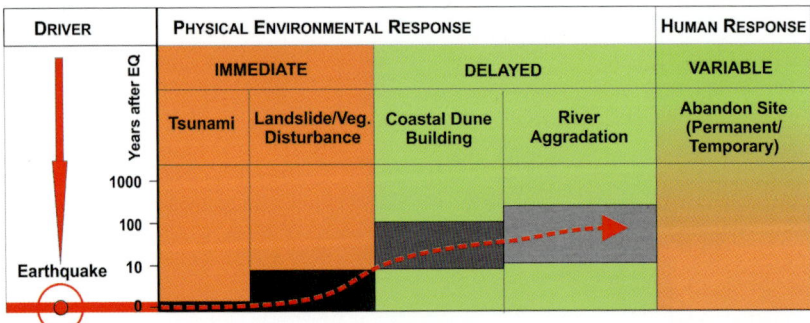

Plate 6.1 Seismic Driving/Seismic Staircase concept. The earthquake driver at 0 years, sets off a series of environment and human response after-effects over different time scales. (Image: J. Goff)

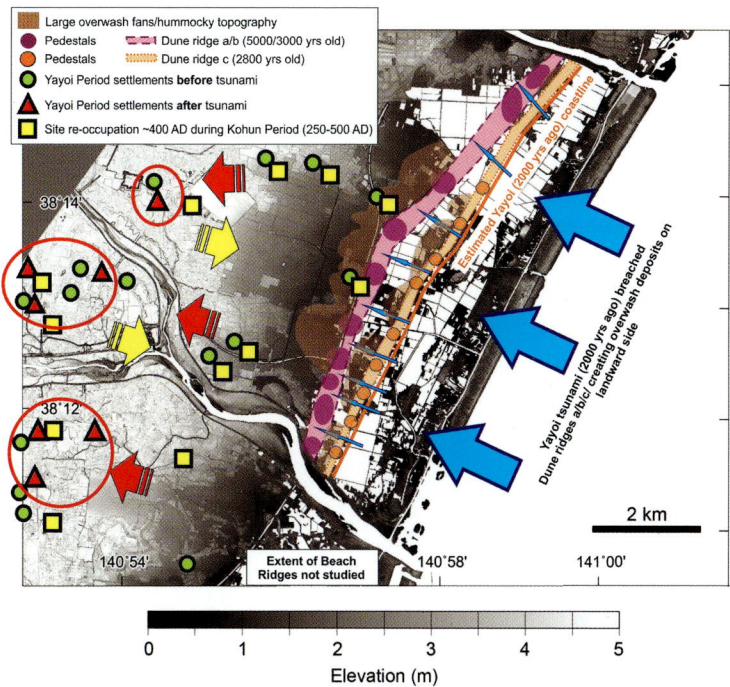

Plate 6.2 Sendai Plain: Purple and orange symbols relate to dune ridges formed after earlier earthquakes (5000, 3000, and 2800 years ago), the Yayoi tsunami (blue arrows) breached these dunes creating overwash fans/hummocky togography (brown shading) and inundating coastal settlements (green circles). People moved inland (red triangles/arrows) only to return to previous low lying sites in the Kohun Period (yellow squares/arrows). Image: J. Goff (LiDAR basemap provided by Dr. Daisuke Sugawara, Tōhoku University, Sendai).

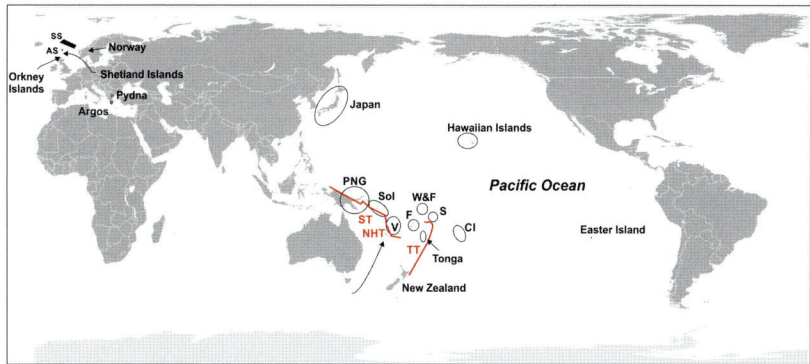

Plate 8.1 World map showing the wide variety of sites talked about in this chapter—and it may be useful for others key sites mentioned in the text at various times. Red lines are subduction zone (NHT = New Hebrides trench; ST = Solomons trench; TT = Tonga trench), Black shaded areas are submarine landslides (AS = Afen slide; SS = Storegga slide), Open black circles and ellipses are places, countries, islands and island archipelagos (CI = Cook Islands; F= Fiji; PNG = Papua New Guinea; S = Samoa; SoI = Solomon Islands; V = Vanuatu; W&F = Wallis and Futuna). (Image: J. Goff)

Plate 9.1 Maullín, Chile—A layer cake of tsunami deposits separated by peat. The uppermost one relates to the 1960 tsunami. (Image: J. Goff)

Plate 11.1 Mejillones Peninsula, the seismic sticking point and home to an unexpected array of tsunami deposits—looking north from Isla Santa Maria (not the one where possible Polynesian skeletons were found, that one is just SW of Concepción in southern Chile, this one is just off Caleta Errazuriz which appeared in Chapter 9—this island has great paleotsunami potential, but that is another journey that is yet to be taken). (Image: J. Goff)

It is actually worse than it sounds. Rivers simply can't cope with such massive amounts of sediment. True, they will shift a lot of it, but a heck of a lot just gets dumped, usually spread out over the land during big floods. But when the floodwaters recede, then all the rest of the sediment just sits in the river channel waiting for the next big flood to shunt it a little further downstream. On occasions though the sediment fights back, and the river shoots itself in the foot so to speak. Sometimes, when the flood waters drop and the river dumps its remaining huge load of sediment in its channel, it suddenly can't flow down that channel anymore. But it has to keep flowing, and so it finds another route and heads off downstream in another direction—this is called avulsion. Fortunately, the Haast River is pretty much confined between steep bedrock valley sides and can't do that, so it struggles bravely on. It really has no option, but we will come back to avulsion a little later on—it is quite terrifying.

So now we have aggradation surfaces. Not surprisingly, these take quite a long time to get down the river to a place where they get deposited, moving much more slowly than the sand. And now we have the perfect sequence of events from start to finish. Unlike Peru, though, where they had high-altitude time-lapse images to show the sequence of landscape changes for their historical event, these prehistoric happenings on the West Coast had forests to chart this ongoing process. With forests come trees, and with trees comes dendrochronology, that dating technique we came across earlier. This technique becomes incredibly important for patching together a chronology of what went on. It is a technique that in many ways proves that all these different adjustments in the environment and people's lives are linked, and if you can't count tree rings to get a really precise age, then you can always use radiocarbon dating to date the wood.

Radiocarbon dating can be used to work out the age of something containing organic material, so wood is good. Radiocarbon is a radioactive isotope of carbon that is constantly being created in the Earth's atmosphere by the interaction of cosmic rays with atmospheric nitrogen. This carbon then combines with atmospheric oxygen to form radioactive carbon dioxide, and as we know, carbon dioxide gets taken in by plants by photosynthesis. If we eat the plants, then we take in the radiocarbon, or we can take it in from an animal that has eaten a plant (cows for example), and in doing so we acquire radiocarbon from many sources. When the animal, plant, or person dies, then this process of acquisition comes to a halt. At this point the amount of radiocarbon in the bones/wood/shell/whatever starts to decrease by radioactive decay. If you work through the logic of this, it is obvious that the older the thing is that you are trying to date the less radiocarbon there is to be detected. Add to this the fact that we know that the half-life (the time taken for the amount of radiocarbon to reduce to half of its initial value) is about 5730 years, then we can realistically work out the age of stuff up to about 50,000 years old. After that there isn't much radiocarbon left, and it is very difficult to detect using the hi-tech instruments we have today.

Needless to say, there are some wrinkles in this technique. It is all too easy to gloss over some of the issues surrounding radiocarbon dating, and while I do not propose to go on a lengthy tangent on the topic, it is important to realize that you do not get a precise date. What you get is an age range that is calibrated based upon known past concentrations of radiocarbon at a particular date, and these known concentrations vary a bit so you end up with a range, say 500–580 years BP (Before Present, which for this technique is 1950 because after that date humans generally started confusing the issue with their nuclear testing, although, as ever, it is a tad more complicated than that!). What this all means statistically (ugh—don't worry, we are not going into any detail) is that the date of the object in question could be any year within that range, although many not particularly enlightened scientists sadly prefer to take a mid-point, thus giving a spuriously precise date—in this example that would be 540 years ago. As is becoming apparent, there are numerous potential pitfalls in the radiocarbon dating technique. For example, the concentration of radiocarbon in the atmosphere has not been constant throughout the past 50,000 years, so working out a date range can be problematic. Marine material (shellfish, seaweed) contains older radiocarbon than terrestrial stuff because it takes longer for the isotope to spread out evenly throughout the oceans. This means that a marine "fudge factor" needs to be added to get to something close to the "real" date, but again that is a bit of an issue since this marine "fudge factor" varies around the planet. Oh yes, and there is a really fascinating point that is well worth reading about. If a human (or a bird or an animal) that died say 2000 years ago had a diet that was dominated by seafood, then using radiocarbon dating, they will appear to have died long before that, often thousands of years earlier. And then a final flag on the play—if you follow this same example, what if they had a mixed diet of marine and terrestrial foods? Ah, well, then there will be a mix of radiocarbon in their system—oh sod it. There are ways of telling how mixed the signal is though, but given all of the ifs and buts related to this dating technique it is always worth taking what someone says about a radio-carbon date with a pinch of salt until you know more of the details.

I used to be driven mad by all of those apparently helpful visitor informa-tion plaques telling you useful things about how old something is. For ex-ample, something like "This archaeological site has been radiocarbon dated to 5000 years ago when . . ." NO, NO, it has not. That is impossible. While I understand that it is easier to say this than something like "This archaeolog-ical site has been dated to between 4800 and 5200 years ago," I find it not only sloppy but a little insulting to my intelligence. Maybe I am just a grumpy old man, and I am by no means blaming my archaeological colleagues for this. It is probably merely someone trying to make a user-friendly sign and cutting out too much scientific detail, or am I meant to think that whoever wrote it thought that most people who read it couldn't cope with an age range or would not really care? Or perhaps the age range is irrelevant to the rest of the story, so why bother? Whatever the reason, this is how BIG mistakes start to get made in science.

Imagine if you had three archaeological sites (I could choose geological sites such as tsunami-related ones, but I am just continuing this theme) within a mile of each other radiocarbon dated to 4800–5200 years ago, 4500–5100 years ago, and 5000–5400 years ago—pretty reasonable ranges to get from radiocarbon dating. OK, so were they all in existence at the same time? The simple answer is—we cannot say, but it is possible that they all were and so this needs to be taken into account, since statistically they could all be 5000 years old. On the other hand, if you took a mid-point (yuck) then you would have 5000, 4800, and 5200 years ago so they couldn't possibly have been around at the same time. Now that is a dangerous approach because it immediately biases your thinking and interpretation.

As the great geologist Thomas Chrowder Chamberlin wrote in AD 1890 in his paper "The Method of Multiple Working Hypotheses": "it is the habit of some to hastily conjure up an explanation for every new phenomenon that presents itself. Interpretation rushes to the forefront as the chief obligation pressing upon the putative wise man. Laudable as the effort at explanation is in itself, it is to be condemned when it runs before a serious inquiry into the phenomenon itself." Strong words but very true—all possible interpretations need to be considered for the case of these archaeological sites—and I should just say now that for these three archaeological sites, I have only mentioned two of the possible interpretations!

Now, back to dendrochronology and counting tree rings.

When a big Alpine Fault earthquake happened, trees were bowled over by the subsequent landslides and their remains can often be found buried in the slides. When a new dune ridge formed at the coast, it was quickly populated by native trees. When an aggradation surface formed, it both buried trees and provided a new surface for others to start growing on. Tree rings have been counted, radiocarbon dates (as back up) have been recorded, and we now know in great detail the dates of all of the Seismic Driving components for the past four big Alpine Fault earthquakes. Why only four? Well, we know a lot more dates for earlier events but tying together all of the component parts—all of the environmental and human responses—can only go back so far in New Zealand since Māori only arrived there about 700 years ago. So, the full suite of Seismic Driving components only works for the last four events with the earliest being in the 15th century.

Let us please not forget the tsunamis, they are after all what this book is about. What on earth has this got to do with them? This is an interesting point in many ways. For example, for the penultimate Alpine Fault earthquake that probably occurred in AD 1717 (a lot of faith is placed in this date, but then some age ranges come out between about AD 1690 and AD 1725, so it isn't carved in stone). We have tree ring data that shows landslides happening in AD 1721, a new Haast dune formed (and stabilized) by AD 1737 (Figure 6.1), and river aggradation (stabilized) in AD 1762. BUT, we have not found the tsunami. We know it must have happened but so far, no luck. But to be fair, it is a very dynamic coast and so the evidence may have been destroyed, or it might simply have been very small and left no visible record.

Now I called the AD 1717 earthquake the penultimate one, and that will have slightly annoyed some of my colleagues, because that was undoubtedly the most recent big one. BUT in AD 1826 (we have already visited that tsunami earlier on in Chapter 4) there was a large earthquake in the Haast region that may, or may not, have been a relatively small Alpine Fault rupture just in that neck of the woods. Whatever fault it was that ruptured, it generated a tsunami, as we have seen. There were also landslides dated to AD 1842 (these have stabilized and new trees have been dated on the old landslide scars), a new dune that formed at Haast in AD 1872, and river aggradation in AD 1889.

Where is the human element in all this then?

Well, I guess I was a bit disingenuous there—we don't have anything for AD 1717, but we do have some serious action on the other side of the country for the earliest event in the 15th century and I will get to that in a minute. And as for AD 1826, we have some rather neat data that was also mentioned earlier.

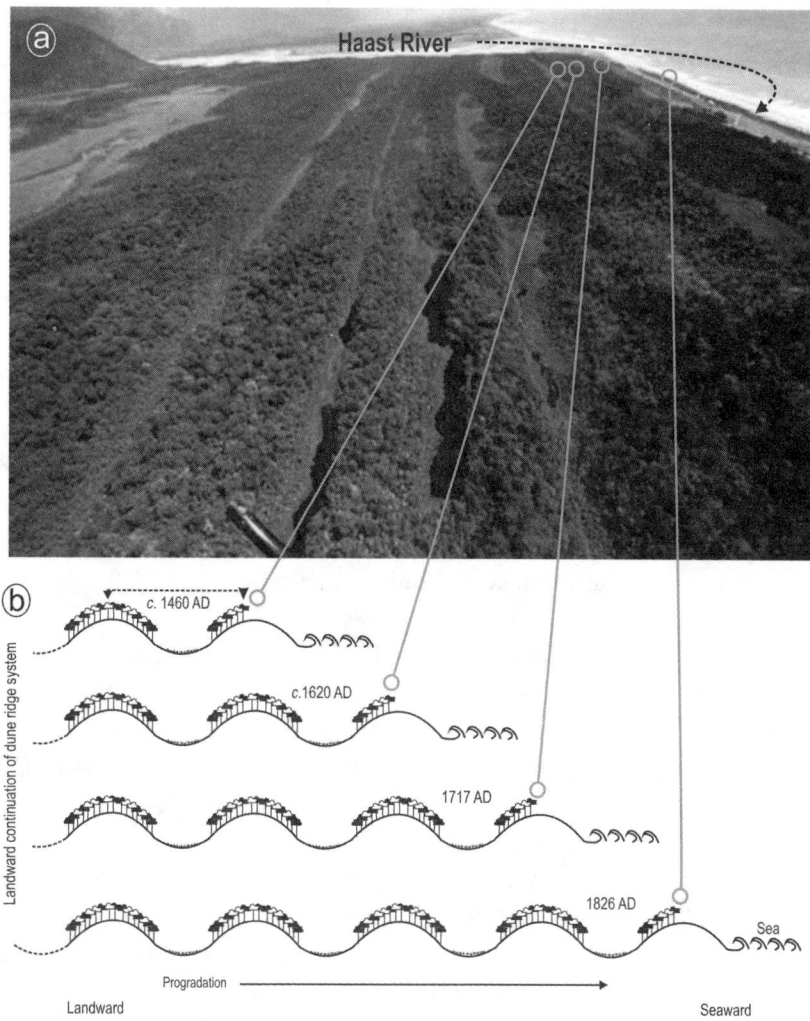

Figure 6.1 Haast dune system: (a) Aerial photo looking south showing the Haast River in the background and Haast dune system in the foreground—the dashed black arrow shows direction of longshore drift that delivers the sand to the coast (Photo: Dr. Andrew Wells; used with permission); (b) Schematic of dune progradation—when a new dune forms trees are able to start growing on both the seaward side of the earlier dune and on the landward side of the new one (after Wells and Goff, 2006). Grey lines link dated dune ridges between the schematic diagram and the photo—many older, undated dunes are also visible in the photo.

The first European to travel down the coastline and comprehensively explore that remote area noted two things in late October AD 1847, 21 years after the event: "the wreck of a large sealing boat amongst a lot of underbrush. It is about a quarter of a mile from high water, and the growth of the bushes and the appearance of the wreck show that the sea is fast receding from this coast." AND at Ōkārito Lagoon, "The timber here is very small, and appears of recent growth. I think to the foot of the mountain range has been recently washed by the ocean."

AND as if that was not enough, another early adventurer and surveyor, Arthur Dudley Dobson, spent seven months in AD 1864 doing survey work along the still almost completely unexplored West Coast. In his book, *Reminiscences of Arthur Dudley Dobson*, he states that, "At the Poherua lagoon [about 15 km NE of Ōkārito Lagoon] there had been at one time a Maori village of considerable size, the stumps of the posts of the houses showing plainly, also the places paved with stones where the fires had been made in the huts. The land had sunk only comparatively lately, as these traces could be seen at dead low water, spring tides."

Ah ha! Remember now, I said that when the Alpine Fault moves (and it would seem that the AD 1826 earthquake did indeed happen in the Alpine Fault system) the land to the west of it subsides and a tsunami comes in. However, we have a tsunami at Ōkārito Lagoon where the land went UP and yet here at Poherua Lagoon, now known as Saltwater Lagoon, we have SUBSIDENCE coupled with the AD 1826 tsunami, and that led to village abandonment and Dobson's observations of what was left. The ups and downs of the coast to the west of the Alpine Fault are indeed a complex puzzle to resolve. We know there are ups and downs, but in any one event it seems that there may be different changes at different places. This is not unique in the world, and similar scenarios have been reported from Papua New Guinea, for example. The big problem is trying to figure out how this happens. It has been suggested that this could be some type of rippling effect that moves up the fault—but the jury is still out on this.

As a minor aside, anyone who has ever been to the South Island of New Zealand will doubtless be aware of the easiest route to cross the Southern Alps, Arthur's Pass. It is well worth the trip, because among other things you will drive across the Otira Viaduct just to the west of Arthur's Pass. Before this was built, the trip over the pass by car was a bit of an adventure as you climbed steeply up and down a narrow switchback road built on one of the

largest unstable landslides I have ever seen! I remember once doing it in torrential rain (nothing unusual there) while being stuck behind the heaviest,
slowest moving truck in the world. I was expecting to plummet to my death
every second of the journey. The viaduct is an engineering masterpiece finished in 1999. It is 440 m long, 35 m above the ground, and has a gradient of
over 11%. There is a great viewing point above it, on the side of the old road
adjacent to what is known somewhat ominously as Death's Corner. If that is
not intimidating enough, then just be warned, if you get out of your car there
to look at the view watch out for Keas.

The Kea (*Nestor notabilis*) is a large, about half a meter (1.5 ft) long, alpine
parrot, olive-green in color with brilliant orange bits under its wings. To say
that Keas are omnivorous is probably a bit of an understatement. They do eat
a lot of the normal food groups—carrion, roots, leaves, berries, and insects.
But they also have a penchant for boots, laces, skis, backpacks, and their most
favorite nibble, the rubber area around car windows! My car used to carry the
scars of one overly peckish Kea.

When you travel across Arthur's Pass, you will be traveling in the footsteps
of some of the earliest Māori settlers as well as early European adventurers.
Arthur Dudley Dobson was one of the latter, and when commissioned to
search for the easiest route across the Southern Alps, he received advice from
the West Coast Māori chief Tarapuhi who told him of a pass that steeply
descended to Otira on the West Coast. Māori had been using it for centuries
to gather pounamu or Greenstone (most New Zealanders wear a bit of this
around their necks) from the West Coast. Arthur reported this find and soon
after that found out that his name had been given to the pass.

Back on the West Coast, we can see from the dates of all the Seismic Driving
components that it seems to take between 20 (AD 1717 to AD 1737) and 46
(AD 1826 to AD 1872) years for a new sand dune to form at the coast following
a large earthquake. This is partly because it rains so much and partly because
the rivers are pretty short. Having said that, it is nothing like as fast as Peru,
but it can also be a lot slower.

On the east coast of the country, across Arthur's Pass, down the eastern
slopes and across the Canterbury Plains is the largest city in the South
Island—Christchurch. The Waimakariri River flows 150 km (90+ miles)
from the heart of the Southern Alps (near Arthur's Pass) to the sea immediately north of Christchurch. It is a massive river that gets most of its
water from snowmelt and glacier runoff, which is useful. If it relied solely on

rainwater it would be a tad smaller, since most of its catchment is in the dry, rain shadow to the east of the Southern Alps divide. But it does have a decent water flow, and that is important because it has to transport sediment far further than its West Coast counterparts.

Things are a bit more complicated on the east coast because many of the Seismic Driving after-effects are attenuated and so, not surprisingly, humans don't necessarily "get it" and instead see each component of the Seismic Driving package as an isolated event. For example, an Alpine Fault earthquake around AD 1250–1300 caused massive amounts of landslides in the Southern Alps (tick that box). Somehow, probably from an associated fault or perhaps a submarine landslide caused by all the shaking, a tsunami inundated the Christchurch coast (Figure 6.2). We know this from a lot of geological work carried out along the coast here, but particularly from Moa Bone Point Cave on the southern side of the large estuary on the seaward edge of Christchurch (Avon-Heathcote Estuary (*Ihutai*)). Driftwood, marine sand, and pebbles were washed into the cave, which acted like a giant washing machine as the high energy flows mixed the marine debris with material that had fallen down from the roof as the ground shook during the earthquake.

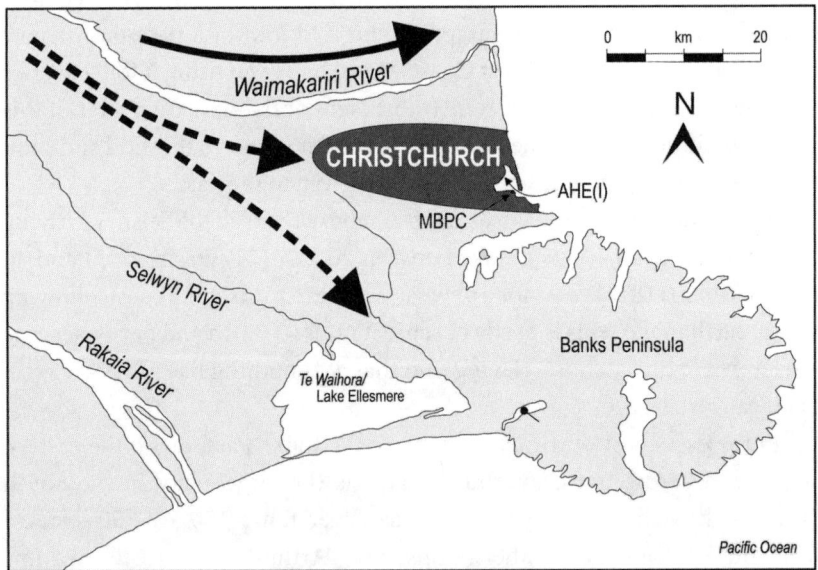

Figure 6.2 Christchurch area: The Waimakariri River currently exits to the north of Christchurch BUT due to avulsion can either exit through the city or via Te Waihora/Lake Ellesmere to the south; MBPC = Moa Bone Point Cave; AHE(I) = Avon-Heathcote Estuary (*Ihutai*). (Image: J. Goff)

That's not all though, it also buried all the remains of Māori occupation left by the first inhabitants of the cave.

But wait, there's more. At the time of this earthquake the Waimakariri River wasn't flowing out to the sea just north of Christchurch, oh no, that is where it is now. When the tsunami inundated Moa Bone Point Cave what is now known as the Avon-Heathcote Estuary (Ihutai) was an open bay. The Waimakariri River was flowing miles to the south and entering the sea via what is now called *Te Waihora*/Lake Ellesmere (on the map Christchurch is just to the north of Banks Peninsula and the lake is to the south). This meant that any sediment being delivered to the coast was coming out to the south. If it had been coming into the sea to the north of Christchurch, then longshore drift to the south would have carried sand down toward the bay to form a spit that would have closed it in to form an estuary—no estuary means the river was flowing to the south.

We also know that the Waimakariri River was flowing south because early Māori used it to travel inland, and they camped at various places along its banks. These archaeological sites now rest on the banks of the old channel. Oh yes, and once the tsunami had inundated the cave and the Māori living there had abandoned it, they moved to a higher cave a short distance away, but then moved back again sometime later . . . because the spit was back in place, and it was safe to live there again.

Hang on, where did the sand come from to form the spit?

Yes, the Waimakariri River had avulsed. So much sediment came down the river following the AD 1250–1300 earthquake that as soon as it got to the flat land inland from Christchurch it dumped most of it—it was too difficult to move it any further downstream. Doing this plugged up the southern channel and so the river avulsed, changing its course to flow north again. This was a two-step process though, it didn't switch across immediately to its current position. It first moved part of the way north and flowed straight through where the middle of Christchurch is today before it avulsed again to its northern exit (Figure 6.2). The Waimakariri River essentially has three choices of where to flow within the confines of its catchment: North (where it is now), Middle (through Christchurch), and South (out through Lake Ellesmere). We know this because it wasn't just this earthquake that caused it to avulse, the same process has happened at least twice in the past few thousand years alone.

Unlike the rapid transport of sediment through the system in Peru and the West Coast of New Zealand, this entire delivery process can take anywhere between about 100 and 300 years.

Some worrying points come out of this work. First, the groundshaking associated with a large Alpine Fault earthquake can generate submarine landslides and/or set off underwater faults near Christchurch. Second, these can generate tsunamis. At the moment, when this happens next the spit now in place will protect the estuary and the people that live around it. Though those who live on the spit won't be protected. Third, a couple of generations later all hell will break loose as a result of this earthquake when the coarse sediment finally gets deposited as an aggradation surface on the Canterbury Plains. The river will change its course again—avulse—and it will go south. The big question here is, will it go to the Middle (Christchurch) or the South, or both as it moves south? Over 380,000 people live in the city, with many more in and around the region that would be affected.

So tsunamis are part of a bigger picture of hell that gets unleashed after a decent earthquake. We have established this from a moderate earthquake in Peru and from larger earthquakes in New Zealand. All have their own unique set of variables, but the components are all there with tsunamis inundating the land and affecting human coastal communities. BUT perhaps this bit about humans being affected seems just a little thin. After all, the evidence for a large number of humans being affected is not great, the tsunamis are a little small, and none of this seems particularly immediate and frightening.

The reasons for going through these examples first though is because this is where the ideas came from, where the dots of human and environmental responses to large earthquakes were joined up, and where many of the little permutations and combinations that cause these to vary from place to place around the world were thrashed out.

Not convinced? OK, let's go to Japan.

In many ways this is the mother lode. In 2011 the disaster everyone was waiting for happened. After all, Japan is recognized as the most tsunami-prepared country in the world.

The 2011 Tōhoku earthquake and tsunami happened just before 3:00 p.m. on the March 11. The magnitude 9.0–9.1 earthquake occurred about 70 km (43 miles) offshore from the Sendai Plain (350 km [about 220 miles] north of Tokyo) in the Japan Trench at the boundary between the Pacific and Eurasian tectonic plates. It was the fourth most powerful earthquake in the world since modern record-keeping began in 1900. The ground shook for SIX minutes.

And the Seismic Driving scenario began.

There was only about a 10 minute warning of the tsunami that was generated by this earthquake. But that was fine, there were plenty of evacuation points to go to and an educated population. The tsunami traveled at around 700 km/hr (430+ miles/hr) and when it inundated the coast it reached run-up heights of up to 40.5 m (133 ft), penetrating up to 10 km (6 miles) inland.

Many people did what they were meant to do and managed to get to local tsunami evacuation sites. Unfortunately, hundreds of these sites were destroyed or completely washed away. They were prepared for a tsunami, but not one this big. Officially, 19,747 deaths have been reported BUT there are also 2,556 people missing.

It is important not to simply gloss over this horrendous tragedy. How was it possible that so many people died? How was it that the most well-prepared country in the world was brought to its knees?

The reason is quite simple—hubris.

In 1952, Japan was well ahead of many other countries in the world and had already established a tsunami warning system to address the problems of tsunamis associated with large earthquakes occurring close to their shores. Like many great ideas though, this had its flaws.

The main problem was that only eight years later their eastern shores were struck by the AD 1960 Chilean tsunami that killed at least 139 people and destroyed well over 1500 houses. There were two incredibly important outcomes from this event. The first was the realization that they were actually exposed to dangerous distantly generated events from across the other side of the Pacific—which they hadn't counted on. This wasn't just a problem for Japan, it was a problem for all countries with a Pacific Ocean coastline—hmmm, international collaboration was needed. The second outcome, which was to have devastating ramifications 51 years later, was that the height of the waves associated with this surprisingly large (and distantly generated) 1960 event was used as the design specifications for the seawall later built around nuclear power plants such as *Fukushima Daiichi*.

At some point in learning all of the lessons from the 1960 event they lost the plot when it came around to preparing for locally generated, and much bigger, events, which had originally been their main concern. The 2011 Tōhoku-oki tsunami was just such an event. When the Fukushima Daiichi plant was built, they worked to a wave height of 3.1 m (10 ft), as opposed to the 14.0 m (46 ft) of run-up that occurred there in 2011. It was a classic case of first taking their eye off the ball and then compounding the error by not even reading their own history. Greater attention should have been paid to evidence from further back in history.

Japan has a deep written history with writings referring to tsunamis as early as AD 684. Perhaps more pertinent is reference to the AD 869 Jōgan tsunami (the previous big one before the 2011 event in Japan) described as "海口哮吼, 聲似 雷霆, 驚濤涌潮, 泝洄漲長, 忽至城下, 去海數十百里" or "sea mouth barking, the sound was like thunder. Turbulent large wave and the welled-up seawater run the river and quickly reached to the castle town that is located far away from the coast." If you marry this written information up with geological research that has been carried out in the region over the last decade or so, you find irrefutable evidence of much larger tsunamis than the 1960 Chilean one. Deposits can be traced kilometers inland, and if recent research is anything to go by (and I hope it is because I was involved in it) the deposits seem to run out about 60% of the way inland. In other words, deposits found 3 km (2 miles) inland show that the water would have gone 5 km (3 miles) inland.

Associated work using geochemistry to trace the saltwater signal and microfossil research to look for marine beasties has now pushed the ability of geologists to trace a tsunami to its maximum inland extent. This is cool stuff but requires skill as well as luck in some ways to ensure that you look in the right place.

One does not even need to imagine the frightening scenario where, with admirable Japanese efficiency, thousands of people rushed to their nearest evacuation point (usually a strongly built, multistory structure, but sometimes simply a footbridge over a road) and waited there in apparent safety only to die as the water overwhelmed or even toppled the structure. One does not have to imagine it because it has been captured on film, CCTV, security cameras, and an almost endless array of live action hell.

Let us step away from the human tragedy for now and wander back to Seismic Driving. If the model is to work then the massive underwater earthquake must have generated a large tsunami—yes, we have covered that, the world knows that. Few know the rest of the story though, much of which is still unfolding, but evidence from previous events show what is underway. Along with the tsunami we have the other immediate after-effect of the earthquake to consider—the landslide component.

The 2011 earthquake generated a huge response from international researchers and the data gathered has been staggering—some is still being analyzed as I type. The sole aim of one set of researchers was to study landslides. So we know right away that there must have been some and that they were expected, because scientists rushed in to check them out. In this case the results are awe inspiring. They identified 3477 large landslides across an area that was up to 200 km (125 miles) from the epicenter. They didn't look any further, there was only so much time, but it is known that there were some landslides as much as 500 km (310 miles) away.

We need to think about that for a second. The earthquake occurred 70 km (43 miles) off the east coast of the Sendai region. At this latitude Japan is about 130 km (80 miles) wide from east to west with the mid-point between the two coasts being formed by the peaks of a large mountain range. This mountain range forms the headwaters of rivers that flow to both the east and the west coasts, so an earthquake off the east coast of Japan caused landslides across the entire country. There was sediment on the move everywhere.

Back to the Sendai area—Sendai city and the Sendai Plain. The Sendai area has been inhabited for thousands of years, with people mainly occupying the fertile land on the plains with easy access to the sea. The history of Sendai city though starts in AD 1600, when the daimyō (feudal lord) Date Masamune relocated there. The city is located about 12 km (7.5 miles) from the sea that lies to the east on the other side of the Sendai Plain. The Sendai area is well-known in Japan for producing excellent sashimi, sushi, and sake—I can attest to that! This is because it has several major fishing ports AND the Sendai Plains is a major producer of fine quality rice, usually two crops a year.

The 2011 tsunami devastated the rice industry, not only covering many of the paddy fields in sediment but also inundating the plains with salt-water. Here the tsunami traveled 5 km (3 miles) inland, so while the city was safe, rice production crashed. An indication of the importance of rice to the Sendai economy was the rapid and sustained remedial action carried out to try and get it back on its feet again. I witnessed this incredible effort firsthand in 2012 while we were studying not only the 2011 tsunami deposit but its numerous precursors. Our team was in one paddy field taking samples and photographs and doing everything that geologists need to do, while in the neighboring field the bulldozers and mechanical diggers were literally scraping the 2011 tsunami deposit off the surface and taking it away to a massive dump up in the mountains. Once cleared, the paddy fields were filled and refilled with fresh water to flush out the salt, repeating the process until it was gone. The scale of this operation was astounding, but it worked. Rice production was only halted for two years which, while still devastating to the local economy would have been nothing compared to the damage caused by previous tsunamis when such modern technological help was not available.

The intensity of the post-tsunami recovery effort was truly amazing. Vast piles of debris were sorted for recycling—wood, metal, plastic, and so on—this was not simply a case of picking it all up and throwing it into a hole to forget about. And once all of the big debris had been cleared, there was still a seemingly endless array of detritus. Bottles, cans, bits of plastic, paper, all the small little bits of life that we tend to ignore until that life of thousands is scattered across the landscape. It might have been a daunting task to clear up, or indeed it could quite easily have been left to be gradually taken over by grass and new buildings as time ticked along. But no, at weekends people would take the train up from Tokyo and form teams who, on their hands and knees, would progress across paddy field after paddy field, tweezers and bags in hand, picking up everything, absolutely everything, that should not have been there. It was an amazing sight, a national effort of monumental proportions, one I glimpsed out of the car window during a torrential downpour—the team I watched carried on doggedly, pursuing their goal of clearing the paddy field.

While Sendai city was safe, many of its communication routes were cut, either by the tsunami or landslides. The earthquake, landslides, and tsunami happened in March—it is cold then, very cold, and this was yet another major problem for survivors to endure, but fortunately there was little rain. The real rain starts in the typhoon season that extends from about May to October, so there was a little time to wait before the fresh landslide-deposited sediments in the upper catchments of the rivers got a taste of what was to come. Typhoon Roke (September 20–21, 2011) alone produced 330 mm (13 inches) of rainfall in the Sendai area. This set the next phase of the Seismic Driving process into operation. And here we hit a wrinkle.

Critical to the functioning of the entire Seismic Driving scenario is that there is a through-flow of sediment within the river systems. In other words, the fine sediments sitting in the upper catchments of the rivers are free to be transported all the way to the coast. A distinctive geomorphological expression of this seismically driven connectivity would be something like the Haast dune system—a prograding coast (one that advances seaward as more sediment is supplied to it) that contains sand dune beach ridges. As we have seen, it may take decades or even centuries for these beach ridges to be formed by this process. This time lag represents the timeframe of sediment transport pathways from source to sink along river systems.

Prior to the 2011 event, major historical earthquakes struck the region in AD 1611 (the Keicho earthquake, a bit smaller than the 2011) and AD 869

(Jōgan—we have heard about this earlier). Both generated tsunamis that can be geologically traced over at least 100–150 km (60–90 miles) of coastline here. We know that the resulting beach ridge from the Jōgan earthquake started to form on the Sendai Plain immediately before a significant volcanic eruption deposited an ash called the Towada-a tephra in AD 915 on top of it. In other words, the ridge started to form about 40 years after the earthquake.

There are many beach ridges on the Sendai Plain that extend well back into prehistory. By studying the dates of these and matching them up with their associated earthquakes it appears that the journey for sediment of around 50 km (30 miles) from the mountains to the coast across the Sendai Plain is completed in 20–100 years. This variability is largely related to the longevity and intensity of rainfall activity. Fascinatingly, this timing fits rather well with the two examples I used from New Zealand.

The wrinkle? What about the wrinkle? Ah yes—humans. We do like to do our bit to screw around with Mother Nature.

Between the 2011 earthquake and the previously large local event in AD 1611 (Keicho) a considerable amount of anthropogenic activity has occurred throughout the Sendai region, and in particular along the rivers and coastline. There has been major river engineering in the form of dams and the coast is now replete with numerous harbors and jetties. Dams on the Abukuma, Natori, and Nanakita rivers, the three key rivers that drain the Sendai Plains into Sendai Bay, now act as excellent sediment traps, stopping the sediment from reaching the coast and ultimately blocking the formation of the 2011 earthquake-related beach ridge. These dams will not only stop the new beach ridge from forming but when coupled with the building of ports and jetties, these obstructions have led to accelerated coastal erosion over decades. The dams stop the sediment from getting down the rivers, and ports, jetties and other coastal structures impede longshore drift, so that any sediment that does make its way down the rivers to the coast doesn't get very far once it arrives.

This is not exactly forward thinking by coastal engineers. Or perhaps it is.

Based upon what happened following earlier earthquakes it seems that in the normal run of things, sand delivered to the coast following the 2011 earthquake should have led to the Sendai coastline eventually prograding (moving seaward) by as much as 1 km (0.6 miles) further out to sea, with at least one new beach ridge forming before the next big earthquake and tsunami strikes. This would have continued the growth of an increasingly effective natural tsunami barrier, pushing the coast further away from exposed residential, agricultural, and industrial infrastructure. Of course, this assumes that the land would not have been built on!

So what has happened to remedy the problem?

An approximately 30 km (19 miles) long, 7.2 m (23 ft) high seawall has now been built along the entire Sendai Bay coastline. This serves two purposes, the most obvious one is that, in a way the 2011 beach ridge has already formed, it is an "anthropogenic beach ridge." A more cynical view would be to look at this from an engineering perspective. Here we see the continued propagation of engineering errors with the construction of dams and ports leading to larger and more extreme structures being built along the coast in order to provide a relatively long-term solution. This will see positive ongoing job prospects for engineers since in reality the lifetime of the seawall alone is probably around 50–60 years. This is as opposed to natural, albeit lower, dune ridges that are still preserved in the landscape after thousands of years. However, the new seawall makes people feel safe.

What price for peace of mind? Well, it is not really even worth considering that question because ultimately a seawall was the only effective solution in this engineering-dominated landscape.

It is also worth questioning whether the natural dune ridge barriers are truly that effective. They may only be 2–3 m (6–10 ft) high, and we saw what happened to the *Fukushima Daiichi* nuclear power plant, which had a marginally higher seawall. The only difference between the two is that the dune ridges exist as part of an extensive natural sand plain that pushes the coastline further away from human communities, so the tsunami has time to lose energy as it moves inland. Additionally, people are mobile—as the coastlines moves further out to sea they have more time to evacuate when the next tsunami occurs, or as a longer-term strategy communities have time to relocate to safer ground inland and uphill.

In considering this relocation to safer ground we start to get a really good idea of human folly. While it has been mentioned before in this book, the Sendai Plain is an excellent example of what this means in real life, and death. To get a true sense of the long-term perspective of human folly we need to delve into the field of what the Japanese call *tsunami kokogaku* or tsunami archaeology. To my knowledge no other country has seen archaeologists specifically target this area of human-environment interaction. It is eye opening, well researched, and offers up the occasional genuine facepalm material.

Let us go back some 2000 years or so to the Yayoi Period in Japan and start there (Plate 6.2). Archaeological information tells us that at that time people were mainly living back from the coast behind a beach ridge put there after an earthquake and tsunami about 2800 years ago. There were fairly dispersed across the fertile plains. Life was good. And then 2000 years ago the Yayoi tsunami struck, after the earthquake. Just like in 2011, the low-lying plains were inundated and immediately following the tsunami we see the two classic responses in the Yayoi Period. Communities moved both inland and uphill, focusing settlement in fewer communities.

Perfect—good move. While people undoubtedly died at the time, the long-term safety of the human population was ensured by this appropriate response.

Nope, it wasn't.

Within 400 years, during the Kohun Period, many communities had moved back on to the plains. In many cases they reoccupied the same locations again, while some also stayed back in the hills as they had before. This is understandable at one level—less distance to travel to your boats and the paddy fields. At another level, it shows that there was a lack of understanding of what life was like in that environment or at the very least a loss of memory of the earlier event or of the importance of that memory.

Life was good again for another 400 years or so though, until AD 869 when the Jōgan tsunami struck, and the process was repeated. The only difference between then and now is that we have this information at our fingertips, recorded by Tsunami Archaeologists. We have an understanding of Seismic Driving. We know what happens in the natural system. Therefore, we should know better and respond accordingly. And so why now after the 2011 event are people once again moving back into the inundation zone? It is because we now have the technology, or at least we think we have, to be able to withstand the next onslaught. It is an interesting situation. Is it hubris or will we vanquish our foe?

My main concern for Japan is that now they have brought the coastline to a point where they need to constantly manage and improve it into the future, will economic pressures see such necessary endeavors lose out to more pressing needs? After all, it may well be a long, multi-generational wait until the next tsunami, and sea walls need replacing . . . often.

As George Santayana (1863–1952), a famous 20th century American philosopher wrote in his 1905 book *The Life of Reason: Reason in Common Sense*: "Those who cannot remember the past are condemned to repeat it."

Enough said.

7

Strand 6

The Human Touch

In addition to the foregoing, a carbonised coconut was found, together
with a fragment of a human skull. A search for teeth or other relics was
conducted without success.

> J. Nason-Jones. Notes on the Geology of the Finsch Coast Area,
> northwest New Guinea.

We now delve more deeply into the prehistory of the human side of the story.
That is where we need to go. I have mentioned it before, but if it wasn't for
us, tsunamis would just be part and parcel of what goes on in the world. We
have an annoying habit of getting in the way of tsunamis. We don't like it, so
we try to avoid them or at the very least mitigate against them. It is amazing,
though, how we do indeed seem to quite spectacularly fail to learn from either history, archaeology, geology, or some combination of all of them.

I have talked about New Zealand and its tsunamiscape and the abandonment of coastal settlements following a nasty event (or events) in the 15th
century, but that has all been rather abstract. I haven't really delved into this
part of the story in any great detail. So as we venture more into the human
elements, it is important to recognize that the modern era is not exceptional.
Just remember the widespread destruction that occurred during the 2004
Indian Ocean and 2011 Japan tsunami events and now go back in time.

While tsunami archaeology has been grabbed with open arms by Japanese
scientists eager to understand more about the human element of past
tsunamis, this has not been the case elsewhere. Many archaeologists still
cling to long-held beliefs to explain some of their unusual finds in coastal
settlements, or at the very least they find comfort in explaining what they
find within their existing research paradigm (or underlying world view). For
others, tsunami archaeology is a breath of fresh air, albeit rarefied.

In Search of Ancient Tsunamis. James Goff, Oxford University Press. © Oxford University Press 2023.
DOI: 10.1093/oso/9780197675984.003.0007

Let us examine this in a little more detail. I make no apologies for using another example from New Zealand, but other countries make appearances later on, and to be honest I have to choose from places where the work has been done. And so we come to a place called Waitore.

In a sense this is a story of one paradigm meeting another, one the prevailing view of the establishment, the other a new kid on the block.

The Waitore archaeological site is in a swamp on the western side of the North Island of New Zealand. It sits on the southern side of the nose of land that sticks out into the Tasman Sea about mid-way down the island, under the gaze of the moderately volcanically dormant Mt. Taranaki. I use the term moderately because its last major eruption occurred around AD 1655, and it seems to pop off in varying degrees of severity every 100–500 years. So is that dormant or active? Your choice. From a tsunami point of view Waitore is on the side of the country most distant from all the subduction action that is going on between the Australian and Pacific Plates. So since that isn't a subduction source for any tsunamis here, the area has generally been overlooked from a tsunami point of view. That's unfortunate because tsunamis come from other sources too. For example, the area immediately offshore from Waitore is known as the South Taranaki Bight and is riddled with faults. Some of these can generate tsunamis, and undoubtedly have.

There are actually several possible tsunami sources and as such it would be entirely reasonable to think that the coast may have been struck by one or more tsunamis at some point in the past. More to the point perhaps is the plethora of information that suggests exactly that, within 100 km of Waitore, numerous sites provide a variety of evidence. We have remobilized sand dunes, with pedestals, hummocky topography, and parabolic dunes (tick all those boxes), marine silts, sands and gravels either washed into wetlands or overlying prehistoric Māori occupation sites, and the all-important *pūrākau*, five of them at least that I know about. I will not go into the detail of all of these, but the data are out there, published for all to see and read. And so, by a somewhat convoluted and head scratching exercise the date for the only known tsunami to have inundated this area has been put at somewhere between AD 1470 and AD 1510.

Waitore Swamp is a very important archaeological site. It represents the oldest collection of wooden artefacts ever found in New Zealand—these were found in the swamp. The collection is extremely significant in that it appears to be a single contemporary assemblage, and not just wooden ones. Over 100 artefacts have been recovered, including parts of an outrigger canoe (some intact but most broken), a box lid, a canoe bailer, pieces of gourd, a large anchor stone, vine lashings, oven stones, wooden planks and burnt timber, some showing signs of cutting or working, a whale vertebra and fish bones, some lying directly beneath the anchor stone. All in all, a chaotic assemblage that seems to be best described as a snapshot of life on the littoral divide.

Quite logically, a series of interpretations for this assemblage have been put forward based largely upon findings from inland sites. The broken canoe parts were considered most likely to have been thrown away or broken deliberately because of someone's bad temper or some ritual or perhaps warfare. The intact canoe parts on the other hand were placed into the wetland to preserve them. The burnt timber was firewood. In summary, this wide array of artefacts represented a mixture of rubbish disposal and intentional burial, with the odd ritual thrown in.

This swamp clearly served many purposes.

Let's look at these interpretations. It is known, for example, that contemporary Māori carvers often keep their timber underwater to reduce the likelihood of cracking and warping. Using such practices as a guide, it was inferred that the wood found at Waitore was buried temporarily for seasonal storage, concealment from enemies, curing and seasoning, softening and preservation. However, while these are perfectly legitimate explanations, it is not easy to crowbar large anchor stones, whale vertebra, and fish bones into these interpretations. And also—why was the canoe in pieces, and what about the burnt firewood?

While a temporary storage interpretation sits firmly within the existing paradigm, and certainly the paradigm at the time of the dig in the 1970s, it was unable to incorporate elements of the physical environmental that were simply not well understood at the time. This is where two things come into consideration—the growth of the relatively young subdiscipline known as landscape archaeology and tsunamis.

It is a little disingenuous to call landscape archaeology a young subdiscipline, hence the qualifier "relatively" was plonked in front of it. From an archaeological point of view it is young, though, with its roots going back "only" as far as the 18th century to William Stukeley.

Stukeley was a Fellow of the Royal Society (the oldest national scientific institution in the world) and Fellow of the Society of Antiquaries of London. He was an Anglican clergyman, a physician, AND an antiquarian—a member of that old school of gentlemen scientists. An antiquarian (Latin: *antiquaries*, meaning pertaining to ancient times) essentially studied things of the past—ancient artefacts, archaeological sites, manuscripts, and so on. These people were actually trying to be scientists. They didn't want to waffle on with fanciful theories, but rather wanted hard facts to work with. Stukeley's pioneering works included studies of the ancient sites of Stonehenge and Avebury in England and Ephesus in Turkey. Among his many findings, he noted that in earthquake-prone areas the shocks were more violent on rocky ground, and buildings on boggy ground suffered from liquefaction of the ground. These observations of human-environment interactions marked the very beginnings of landscape archaeology.

The value of landscape archaeology is that it requires a skillset that lies outside that of the archaeological dig, outside that which has underpinned archaeological thinking for centuries. In many ways it is very close to tsunami research because it is multidisciplinary and encompasses a range of expertise from geomorphology to anthropogenic activity to the cultural considerations of social geography. In undertaking such research, its practioners recognize that by not considering such things as landscape and natural processes, the interpretative work of field archaeology (based on a site, artefacts, and other physical remains) will always be restricted to conjecture and possibly trapped within existing paradigms. So landscape archaeology really is something of a new kid on the block, which, like tsunami research in a way, is trying to gain greater acceptance within a field of well-established dogma.

A landscape archaeology assessment of the Waitore site raises two rather important questions. First, where are the remains of the prehistoric Māori village associated with these hidden/stored/ritualistic artefacts? Second, where have the dunes gone? Waitore swamp sits at the seaward end of a very short stream that empties into the sea at a site that is definitely not conducive to the launching of canoes (Figure 7.1). Hmmmm—now that is a rather important point if you have canoe parts for a canoe.

Rewind the field archaeology interpretation and now put a landscape archaeology slant on things.

No village, no dune—how come? Without going into a lengthy explanation of how we got there, it goes like this. The site was inundated by a large tsunami in the 15th century—the dating done during the dig effectively confirms the date of the event. Inundation by a tsunami at this time removed the dunes, destroyed the village (villages were usually located just landward of the dunes), and eroded the wetland immediately landward of the dune. The sand, peat, and artefacts were carried inland and tumbled in the washing machine of the waves and then dumped after a short distance in what was at

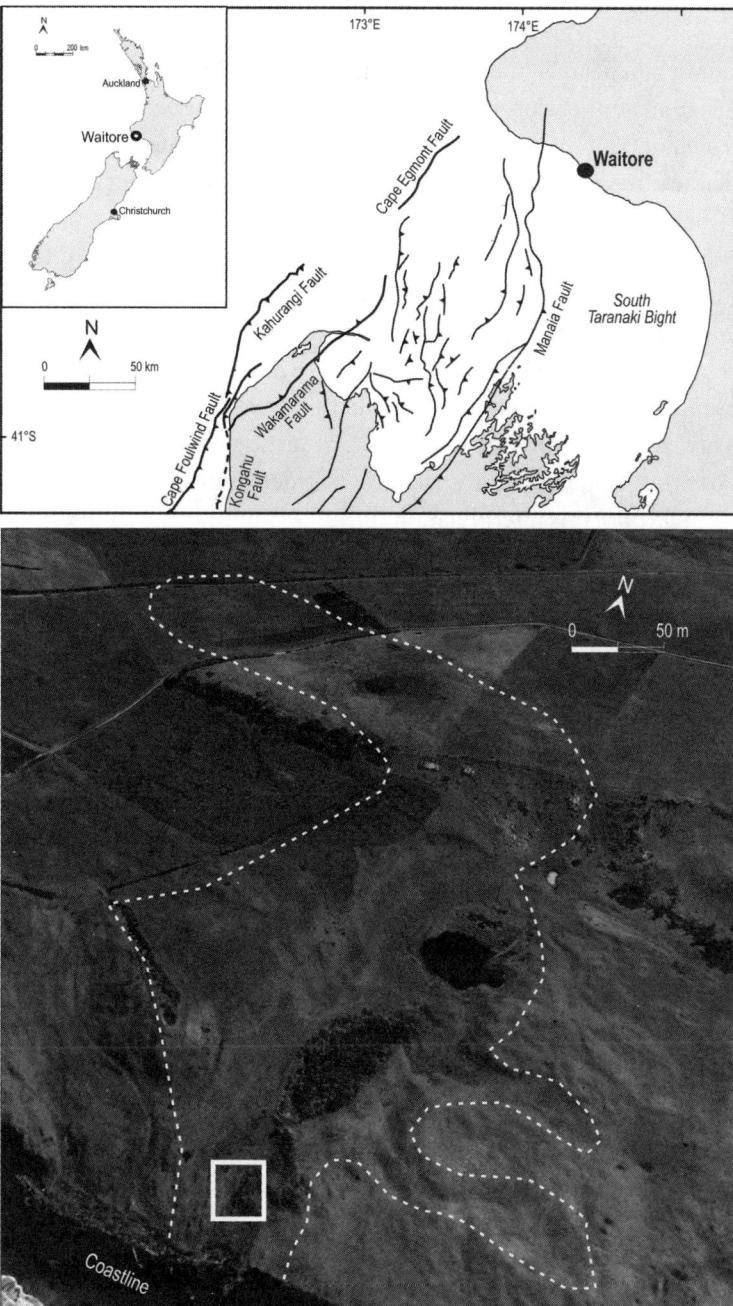

Figure 7.1 Waitore sits on the coast adjacent to a seemingly endless number of underwater faults. The small valley has collected lots of peat and sand from the destruction caused by the tsunami. The main archaeological site is down near the sea (white square). (Upper Image: J. Goff; Lower Image: annotated Google Earth image)

the time a much larger, more substantial wetland. This sudden dumping of a huge load of stuff on top of it seriously upset the underlying wetland and it deformed or bent under the weight of all of this new material. This bending was in the direction of flow, forming what we know as flame structures in the peat (they are visible in the dig photos). While the original archaeological dig identified 11 peat layers (with a sand layer overlying each of them). The radiocarbon dates of the peat layers do not show any differences in age, but peat takes a long time to form, and sand would have had to gradually build up between the layers as well. What seems to have happened then is that the peat layers represent large chunks of wetland that were ripped up by the tsunami and carried inland from the more seaward section of the swamp. The inclusion of other ripped-up material such as older sands and dune vegetation points in favor of this interpretation.

So, based upon this modern take of the data, what we appear to have at Waitore is not a swamp in which things were placed for whatever reason, but an old swamp that got picked up and moved inland and redeposited in a bit of a chaotic jumble. Strike one for tsunamis perhaps!

By using landscape archaeology, researchers seem to have been able to shift the focus beyond existing paradigms and provide a very real physical environmental context in which to place the site. This example shows how well this type of scientific investigation can work at a local level, but its value becomes even more apparent on a regional scale. To do this we could stay in New Zealand, but it is time to start traveling north, and we head back to the French outpost of Wallis and Futuna.

We know from some of my earlier ramblings that the most recent paleotsunami in the Wallis and Futuna archipelago is well placed by radiocarbon dates to around AD 1480. On Futuna, the geological evidence was gleaned from 10 trenches with soil or prehistoric occupation layers overlain by marine sand and coral debris. On the west coast of Wallis there is even more evidence, with the abandoned prehistoric settlement of Utulévé overlain by a marine pebble layer. These events on the two islands match in date and also match the date of other deposits found in New Zealand (there we go again, sorry), Samoa, Vanuatu, and so on. This was a Tonga Trench subduction zone earthquake on steroids that generated the largest tsunami to have struck the region in the past 2000 years or so.

Let us recall what a tsunami does to a coastal community—the parallels between ancient and modern are quite remarkable, so if you do not believe what you are being told about the effects of prehistoric events, you can always check up on recent ones. People move inland and uphill, often permanently abandoning their old settlement, but often this is only a temporary move. Resources are depleted because saltwater inundates the land, the tsunami erodes vast amounts of soil containing crops, and nearshore shell beds are picked up, thrown inland, and destroyed. During the resource crisis, new political alliances are made, and warfare and/or social instability becomes prevalent.

On an island or nationwide basis these changes can be unpicked from the archaeological record, but in the vast island archipelago of the southwest Pacific there are multiple islands and nations, and these human responses can be charted as a ripple effect across the entire region.

For example, after the earthquake and tsunami the archipelago of Wallis and Futuna was invaded by aggressive Tongan raiders who had paddled their war canoes some 700 km (400 miles) northward across the ocean in an opportunistic maritime expansion. While those on Futuna were able to repel them, those on Wallis (the Wallicians) were less successful and the island was occupied. Although some Tongans would undoubtedly have "moved inland and uphill" after the tsunami, not all could. The only way for the Tongans to ensure safety was to adopt a variant—"move on and conquer," and so some of their large canoes must have survived the tsunami.

The after-effects of the tsunami on Wallis and Futuna were therefore two-fold. First, the movement of coastal settlements inland and uphill and second, almost immediately afterward, the construction of forts to protect themselves from invaders. The fact that Wallicians were unable to repel this aggressive Tongan maritime expansion, though, may well explain a unique reversal in the relentless easterly settlement of the Pacific region.

Polynesians only arrived in the vicinity of Tonga around 3000 years ago. Before that all of the islands of the South Pacific were uninhabited, except for those already settled by Tongans along their route from the west, near Papua New Guinea. This eastward island-hopping of the proto-Polynesians or Lapita Culture can be traced through the presence of Lapita pottery on the islands along their route. These proto-Polynesians are known principally because of the remains of their fired pottery, which consisted of beakers, cooking pots, and bowls. Many of the pottery shards found are decorated with geometric designs made by stamping the unfired clay. Lapita pottery has been found all the way from their starting point in Papua New Guinea, southward through New Caledonia and Vanuatu, and eastward to the Western Polynesian island archipelagos of Tonga, Samoa, and Fiji.

I feel here that I must take you off on a tangent. Completely unrelated to Polynesians, but one that involves tsunamis and pottery . . . Roman pottery, Byzantine pottery, Sterile sand, and the Mediterranean coast of Israel.

The place is called Tel Ashkelon, a short hop, step, and a jump from the city of Ashkelon and intimately tied in with it. Tel is a Hebrew word for an ancient mound composed of remains of successive settlements. Most people are familiar with Tel Aviv which is about 50 km (30 miles) north, but Tel Ashkelon is also only about 13 km (8 miles) north of the border with the Gaza Strip, and is a place prone to rocket attacks. The ancient seaport of Ashkelon dates back to Neolithic times, and since then it has had a plethora of owners including the ancient Egyptians, Canaanites, Philistines, Assyrians, Babylonians, Greeks, Phoenicians, Hasmoneans, Romans, Persians, Arabs, and the Crusaders, until AD 1270, when the Mamluk sultan Baybars ordered its destruction.

An important city, Ashkelon was defended on its seaward side by a high natural bluff that is now suffering a considerable amount of coastal erosion, which is also removing evidence of the long history of human occupation. It was this eroding coastal bluff that got the tsunami bloodhounds working. For many years I had wanted to carry out some research around the Mediterranean region, but I was working in the Asia-Pacific and the Mediterranean is generally, and not surprisingly, the preserve of those who live around it, with a few other Northern Hemisphere players from Europe and the United States. As luck would have it, I was fortunate enough to be invited to Israel for a short visiting professorship, and it was suggested one day that we might take a look at the eroding cliffs at Tel Ashkelon. Every storm was taking just a little bit more of the tel, and history was laid bare on the sands below the cliff, gently oozing into the sea to be lost forever.

This part of the world is no stranger to tsunamis and over the past 10–15 years some great work has been done there studying the evidence of their destructive past. We know about several notable events, but I am by no means going to talk about all of them. The earliest documented paleotsunami in the past 10,000 years occurred a little over 9000 years ago. Evidence can be found about 100 km (60 miles) north of Tel Ashkelon at an early Neolithic site, which shows that the tsunami ran-up at least 16 m (50 ft) above sea level and extended inland up to 3.5 km (2 miles). Earlier Neolithic material was completely removed by the tsunami, and the site was abandoned though only temporarily. While the tsunami had been an incredibly disruptive event, this was an important coastal site, even back in those days. A little further to the south of the Neolithic site we come to Caesarea where we find evidence for the infamous Late Bronze Age (1630–1550 BC) tsunami deposits related to the eruption of Santorini. This was yet another Mediterranean-wide event that disrupted life here, but settlement continued. Another tsunami struck Caesarea on December 13, AD 115, generated by a well-known earthquake somewhere near Cyprus that also destroyed Antioch.

So, large events have been slamming into Israel's Mediterranean coast for millennia and yet no one had looked at Tel Ashkelon, which had a continuous record of occupation from Neolithic times, or at least seemed to. This was surprising given that between Caesarea and Tel Ashkelon is Tel Aviv, the economic and technological center of the country. It would be nice to know how exposed it is tsunamis. If a wave struck Caesarea and Tel Ashkelon, this brackets Tel Aviv and puts it very much in the firing line for such events.

Needless to say, we had no research funding and so this was just an exploratory mission. Armed with permission to poke around, myself and my Israeli colleague bravely headed down to the beach to have a look at what was falling into the sea. If we had been archaeologists this would have been termed "rescue archaeology" but as geologists I guess it would have been considered fossicking—a general rummaging around with no particular archaeological rigor.

It is difficult to describe the site because it is an almost overwhelming layer cake of human history. The cylindrical tower of a lined well stood proudly, separate from the cliff like an ancient smoke stack—it had not yet collapsed into the sea, but looked perilously like it would at any time. There was no soil as such or any obvious underlying sediment although it must have been there buried deep under layer upon layer of broken amphorae, pottery, and Roman glass that had slid down the cliff face. The difficulty was that we could not stand far enough back from the cliff to get a true sense of the layers simply because the sea was there, too close for comfort.

After a quick look around we decided to examine the middle section of the cliff, largely because the scree slope of artefacts allowed easy access but also because there were some hints of something up there that was partially obscured by the debris. There seemed to be some layering, not only of potsherds but also possibly of sand. Climbing up there was like crawling up through an Aladdin's Cave of archaeology. Probably all too familiar to archaeologists who work in the region, but for me the variety of pottery and glass was astounding. In most cases I had absolutely no idea what it was, but on occasion a half broken Roman wine glass or Byzantine pot was unmistakable. Amazing.

Standing at the head of the unstable scree slope, so completely up close and personal with the area we had seen from below was quite overwhelming. There were so many broken pieces of amphorae that it was difficult to get any sense of what was going on. We started to clear away all of the debris at the top of the slope, clearing downward to see if there was a base, if there was any structure to this mess.

As we were scraping away, occasionally stopping to admire an artefact that was already en route to a watery encounter with the sea we became aware of some shouting behind us. My colleague and I turned around just in time to see a black rigid inflatable boat (RIB) skittering past us about 100 m (100 yards) offshore. It was full of men in black. Not the movie Men in Black, but balaclava wearing, body-armor-clad soldiers of the Israeli armed forces. Apparently, there was no need to worry my colleague told me, these were just regular patrols that scoured the shoreline checking for any unauthorized landings, or attempted landings. I was fairly convinced that no one could get past that lot since, now that we were aware of their presence, they seemed to be passing with ominous regularity every 10 minutes or so.

Our time window was short so, heads down, bums up, we applied ourselves to the task of clearing off a bit of cliff face. It was not long before a thin sand layer appeared beneath the amphorae. Slowly but surely, we tracked the layer laterally along the shoreline becoming increasingly confident that this was indeed a tsunami deposit. Stepping back after a good hour of scraping we noticed a similar layer a bit higher up and also, in another area yet to be covered by the artefactual scree, there also seemed to be something a lot lower down as well.

By this time we had a closer acquaintance with the sand and amphorae layer we had been clearing away. Not one amphora was intact, which in many ways was useful. We could speculate that some might have been broken up by the tsunami, but then equally this could have just as easily been an old dump site. But no, it could not, because there was a flow structure to their deposition. At first we could not see the wood for the trees but as our worked progressed two distinct patterns emerged, it was as if we were having a set of new glasses made for us as we watched. Not surprisingly as it turns out, the broken pieces of amphora responded just the same way as sediment does when it has been transported and then deposited by water.

Large, elongated cobbles for example will do one of two things. They will either roll at right angles to the flow, just like you might roll a pencil along a tabletop, or they will align themselves parallel to the flow direction, reducing the drag on themselves, just like a plane flying through the air. When they finally stop moving they are often imbricated, like tiles on a roof. This is the most stable position, and they tip back slightly toward the direction of the flow that put them there. This is what we were seeing in the main part of this several-meter-thick layer of broken pots. Because they had all broken in uneven patterns and therefore had jagged edges, it had been difficult to pick this up to start with, other than having vague notion that this was what we could see. It was only when we examined the deposit in detail that it became apparent.

The whizz of Israeli patrol boats was long forgotten.

Perhaps the icing on the cake was almost literally that, it was the topping of the deposit. Once the energy of the tsunami had slowed down and the run-up was nearing its most inland extent, pieces of amphora that were still suspended in the flow started to settle out, not stacked up by the high energy of water but gently coming back to earth. These pieces, unlike their counterparts that had already been laid down, settled out like seashells, not sediment. A disarticulated bivalve such as a mussel has two separate shells, both of which have a convex outside and a concave inside that provides the space to hold the "meat" in the middle. If this bivalve separates into two pieces, and the shells are transported by water and allowed to settle out slowly, they are almost always deposited in a concave-up orientation, like a dead leaf floating down from a tree onto the ground. Some will always be convex-up, though, often simply because they were unable to settle out uninterrupted and got shuffled around in the mix, but it is this general pattern of settling you look for. As opposed to imbrication that is driven by water moving in one direction, here there is no creation of a tile-like form in balance with the flow, instead you have something akin to numerous stacks of plates as each shell (in our case each fragment of amphora) comfortably settles into the divot created by the piece that has just settled beneath it (Figure 7.2).

And there we have the tsunami with high energy creating imbrication leading to a waning flow and the plate-like stacking of gentle settling. In all we figured out that we had one definite tsunami, the one we had been working at all the time, and two others, one younger and one older. But what ages were they? The one we had been looking at was pretty obvious. The artefacts moved were predominantly 1st century Roman and so we were

Figure 7.2 Tel Ashkelon—detail of the cliff section—in the back left is the isolated stack of an ancient well. The base of the tsunami is marked by a dashed white line with the white square showing where some broken amphora fragments have settled out and immediately above some have been rolled and rotated as they were deposited. (Image: J. Goff)

almost certain that we had to be the December 13, AD 115, event that had also struck the coast some 100 km to our north. The one below—perhaps the Santorini event? The one above—not sure, more work needs to be done and that is unlikely to ever happen in my lifetime because we tried several times to get funding but to no avail.

You park such failures in your "I may revisit this in the future" files knowing full well that is unlikely because in all honesty you have to follow the money however small it might be. I imagine that in a few years' time I will see a paper, or perhaps get to review a funding proposal, that aims to do exactly what we wanted to do. And that will be fantastic—it needs doing. Funding is 95% about timing, the other 5% is luck. I had bad timing, again.

Talking about parking though, it was time for us to amble back to the car, happy in the day's work and looking forward to a beer and a chat. We headed back to the beach access and the zigzag of steps to take us back up to the top.

At the first zig we heard a shout from above and looked up to see an almost bowel-loosening sight. Looking down on us from the cliff top were about eight or so balaclava wearing, body-armor-clad soldiers of the Israeli armed forces, with their guns (Tavor X95s, I think) aimed directly at us. Needless to say, my Hebrew is non-existent, and so the shouted commands only served to heighten my level of unease. My colleague smiled, looked at me, and told me to stay where I was—no problem there, I couldn't have moved if I tried. He wandered calmly up the steps talking to them as he went. A quick 30 second huddle at the top and the atmosphere went from tense to laughter. Apparently, those boats zipping up and down the coast had been watching us—we were suspected of having landed illegally and were therefore a possible security threat . . . gulp.

With rubbery legs I climbed to the top of the cliff and with the odd slap on the back that nearly drove my teeth a couple of hundred meters up the road, we parted on friendly terms.

Beer had never tasted so good.

Now it is time to drag ourselves away from this latest tangent, although there is always more to tell.

We are back to the Lapita Pottery making peoples, the proto-Polynesians, who had zipped eastward, island-hopping all the way. This was a rapid movement of people, but it is important to remember that there was no obvious end point to this journey of settlement, it just happened. Why move? No idea, it could have been resource depletion, over-population, the desire to travel, discount rates as a canoe passenger, the list is endless. The point is that they covered a 3500 km (2200 mile) journey of settlement in just over 500 years across an awful lot of ocean littered with islands, many of which they permanently settled, ultimately forming what archaeologists now call Western Polynesia.

This may not seem entirely remarkable today, but it was done in canoes. Yes, ocean-going canoes had sails BUT this journey would have involved a lot of sailing against the wind and as such a lot of paddling. Furthermore, as if it wasn't hard enough, they had to take everything they needed with them AND they could not see the islands ahead of them, they were hidden over the horizon. They had got as far as Papua New Guinea by hopping between visible islands, but this next 3500 km step involved entirely different voyages of adventure. The mere fact that they succeeded speaks volumes about their mastery of seacraft, their navigational abilities, and their intimate knowledge of the maritime environment. Even today in the absence of maps and GPS most of us would be hard pushed and extremely foolhardy to undertake such a journey.

Many books have been written about the Lapita-Polynesian story, and it is a script that is being continuously tweaked and rewritten as new evidence is found. I will visit some of that new evidence later as it relates to an earlier earthquake and tsunami in the southwest Pacific.

For now, it is back to the island of Wallis.

Tongans had invaded and there was a simple choice for Wallicians—accept defeat or run like hell. Running would not have got them too far, the island is only about 15 km (9 miles) long. Paddling though could get them a long way, after all with 3500 km (2200 miles) behind them this culture had the credentials. And so they, or some of them at least, paddled, and doubtless sailed too because they went westward with the wind, backtracking if you wish. This was undoubtedly a pragmatic decision. To go eastward was to battle the winds, which rarely changed direction except during El Niños (there is another book related to that comment too). Not only that but just 360 km (220 miles) to the east was the much larger island nation of Samoa. We know nothing about how well Wallicians got on with Samoans but bearing in mind that only 250 km (155 miles) to the south were the northern islands of Tonga, they were doubtless keen not to run into the more warlike Tongans. Was it worth it? What lay in the opposite direction to the west?

Interestingly, before you eventually end up at a bunch of islands that are now either part of the Solomon Islands or Vanuatu some 1900 km (1200 miles) away, there is simply a vast amount of Pacific Ocean, two solitary islands, and a few reefs. One island, Rotuma, is over 700 km (430 miles) and the other, Anuta, 1500 km (930 miles) away from Wallis. Rotuma is a volcanic island about the same size as Wallis. While the original settlement time of Rotuma is a little unclear, what is known is that sometime after it was first settled, Samoans and then Tongans migrated there. Ah—maybe not a good idea to stop there if you are trying to avoid Tongans then. And so it seems likely that Anuta was the target. Anuta is the smallest permanently occupied Polynesian island in the Pacific, less than 1 km (0.6 miles) long and 0.5 km (0.3 miles) wide—a very, very small target to hit, but they succeeded (Figure 7.3). It currently has a population of about 300. While we do not know what the population of Anuta was 500 years ago, oral traditions on the island indicate that the then local population was supplanted by immigrants from "Uea" (Uvea, or Wallis) who arrived in one or two canoes around AD 1500–1600. What we do know is that immediately prior to their arrival a big,

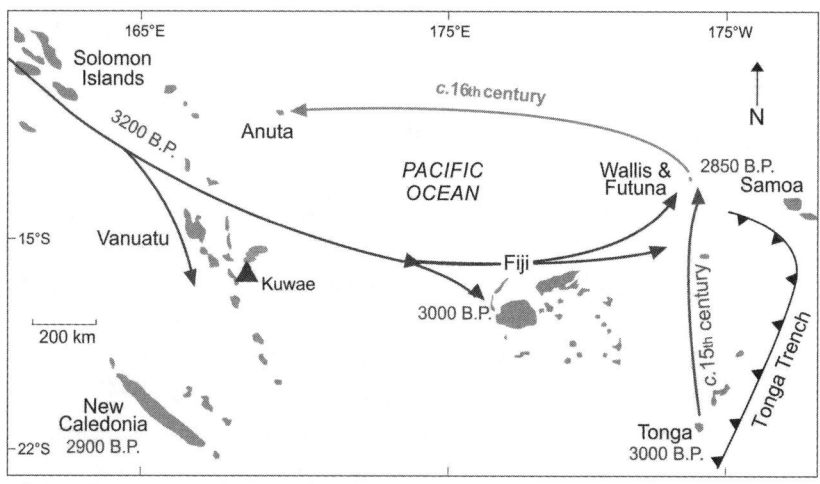

Figure 7.3 The southwest Pacific Ocean. Proto-Polynesians moved eastward from beyond the Solomon Islands arriving around Tonga around 3000 years ago and Wallis and Futuna about 150 years later. Following the large Tongan earthquake and tsunami in the 15th century on the Tonga Trench, Tongans moved north and took over the island of Wallis. A retreat westward saw Wallicians occupy Anuta sometime later. (Image: J. Goff, after Goff et al., 2020)

high energy wave caused a massive new sand dune to form along their coastline. How interesting, we have encountered that scenario before. So here we are again, the human story entwined with the natural environment, two strands of the paleotsunami story both washing up on a distant shore, both telling their own part of the story and this time the story stretched across thousands of kilometers of ocean.

Ocean-wide tsunamis come with ocean-wide evidence. If you thought that joining up the dots to chart the impact of a paleotsunami along one coastline was difficult enough, imagine the amount of work involved in tracking this story down, across hundreds of islands and an ocean that covers one-third of the planet. If you think that is tough, though, imagine trying to persuade colleagues that not only can you trace the evidence for paleotsunamis across these vast distances, you can also chart the human and environmental responses to both the earthquakes (or some other type of tsunami generating mechanism) and the tsunamis across the same area.

It is really an issue of cause and effect. While one may or may not, depending upon who you are talking to, be able to satisfactorily identify tsunami deposits, there is then the issue of dating them and these are very rarely, if ever, precise dates as we have found out. This then leaves researchers with a few dilemmas.

Let's look at a hypothetical example along the Atlantic seaboard of the United States, well away from the Pacific:

- Five years ago Researcher A, a coastal geographer, found a tsunami deposit in the Hudson River by New York—it was dated to about 500 years old according to the media, but is actually somewhere between 400 and 600 years old.
- Two decades earlier Research B, a sedimentologist, had found a deposit related to a high energy marine inundation (HEMI) event (not defined as a tsunami because this was not possible at the time with the knowledge available) at the mouth of Chesapeake Bay some 400 km (250 miles) to the south, and a new group of sand dunes appears to have formed at the coast soon afterward, which lie on top of the HEMI. The high energy event was dated to around 500–800 years ago.
- Researcher C, an archaeologist, had found, some 50 years before that, archaeological evidence for the abandonment of coastal Native American sites on the coast of Maine. This abandonment happened sometime within the past 1000 years, before the written record began.
- Two years ago Researcher D, a paleoecologist, finds in sediment cores taken on the east coast of Nova Scotia what appears to have been an instantaneous change in intertidal shellfish species around 300–500 years ago.
- Finally yesterday, Researcher E, a seismologist, identified a previously unknown undersea fault some 200 km (125 miles) off the coast of Connecticut. This has been traced a short distance along the coast of New England, but while they know it has been active recently, they have no idea whether it was active over the past 1000 years or not.

A selection of just four of the dilemmas this throws up:

a) Did Researcher A really find a tsunami deposit?
b) Was Chesapeake Bay a tsunami deposit as well?
c) Is there a link between all of these events?
d) How on earth can any single researcher expect to find all of this information spread across five different disciplines and 75 years, let alone make any links between them?

Welcome to my world! And please remember—this is a "simple" hypothetical example.

One of the more unsatisfying elements of the entire set of linkages between earthquakes, tsunamis, and humans is the lack of bodies. This may sound a little gruesome, but we know from historic events that many tens to thousands of people can die as a result of tsunami inundation. This has been brought home to us in harrowing and brutal detail particularly since the 2004 Indian Ocean tsunami. So where are the bodies from prehistory?

Now that is a very interesting question, and it requires a little digging to find the answer.

While often reported in immediate post-tsunami surveys, vertebrate remains are rare in older deposits, and there are very few examples of human skeletal remains. Over the past few years some great archaeological and geological work has come out of the east coast of Africa, more specifically from Pangani Bay in Tanzania. The possibility of past tsunamis inundating the east coast of Africa had always been a bit of a bugbear for me. For about the past two decades I have tried and failed to get anyone interested in doing work there, so it is really pleasing to see an international collaboration of scientists working on it now.

First of all these researchers have done a great job at identifying an old tsunami deposit that inundated the coast about 1000 years ago. This has some important implications for the east coast of Africa given that no one knew about this before. It therefore begs the question—how many more are there to be found there? Probably more exciting though is the presence of human skeletons, isolated bones, and bone fragments in the tsunami deposit. These human remains consisted of men, women, and children, many with broken bones. They are randomly scattered throughout the deposit and there is no indication of them having had a traditional burial. There are no associated weapons buried with them, no signs of blunt force trauma or disease, which excludes death by warfare or some type of epidemic and suggests a series of unexpected deaths. Furthermore, the good preservation of the bones also suggests that they hadn't moved too far, and given their association with the tsunami deposit, it is most likely that they drowned.

This is an incredibly rare find, which is rather surprising given that we are generally worried about tsunamis because they kill people and one would assume that there should be some bodies that have not been buried. A brutal statistic from the 2011 Japan tsunami that I have alluded to earlier is that there were 19,747 deaths, 6242 injured, and 2556 people missing. That is

2556 "missing." It is a sobering thought, but from a purely logical perspective, some of those missing must be buried in tsunami-related sediments somewhere. Although tsunami deposits are usually too thin to cover human bodies, in some places they are thick enough to do so. But bodies may have been carried out to sea and covered by sediment, or possibly washed into coastal lakes.

While the discovery of human remains in past tsunami deposits is rare, it is clear that they can be found. I have certainly seen my fair share of bodies, like the hastily buried corpse in a shallow grave on a beach in Sri Lanka following the 2004 Indian Ocean tsunami. While often the indescribable smell of death lingered in the air at many of the sites we visited in Sri Lanka, in this case I was unprepared for the intense sickening mixture of the smell of death coupled with body fluids leaking into my boots when I had the misfortune to stand on the precise spot where the burial had taken place. Enough said. Never forgotten, it merely added another barnacle to the hull of my unpleasant tsunami experiences.

My first encounter with bodies and deaths came following the 1998 Papua New Guinea (PNG) tsunami. While this event probably only affected about 30 km (18 miles) or so of coastline, it obliterated several villages, including Warapu at the end of a spit of land fronting Sissano Lagoon in northern PNG close to the border with West Papua. There had been a moderately large earthquake just offshore, the only real effect of this being to lower the lagoon by about 40 cm (15 inches) and raise the land up a little on the seaward side. This tipping motion in itself would have shoved a little tsunami into the lagoon, but the main issue was that the earthquake generated a large landslide on the steep slopes just offshore. This landslide generated a tsunami with its largest waves and energy focused almost exactly towards Sissano Lagoon.

More than 2000 people died. A wall of sand-laden water rising up to at least 17.5 m (57 ft) high inundated the spit. Virtually no structures were left standing and while many were wooden huts, even the studier buildings had the concrete stripped from them down to the reinforcing metal. Trees were ripped out of the ground and carried more than 1 km (0.6 miles) into the lagoon. The people didn't stand a chance. To the front they had the sea and behind them was the lagoon. If they had run immediately they would have first had to get at least 2 km (1.25 miles) along the spit before they could go inland, or failing that they might have had time to launch their canoes and get far enough out to sea to avoid the full power of the waves. While these were the best options, neither seems likely to have been entirely successful because the first wave arrived only a few minutes after the earthquake.

This was essentially a bad place to live, but, hey, people lived there in the past and they live there again now, although not quite in the same numbers as before. A mixture of necessity (fishing), continuity (we have always lived here), and the ability to downplay the likelihood of another event happening anytime soon is doubtless responsible for this.

It was a surreal experience walking down the old "high street" of the village, a sandy strip of ground. To the left and right were gaps in the palms trees and ironwood stands where houses had once stood. At the end of the village, on the left just before the end of the spit, was the old cemetery, we knew that because we could see all the bodies laid out in neatly dug holes. The only odd thing here was that these were not tsunami victims, they had been long dead. The tsunami had merely removed all of the loose sand from above them and washed it into the lagoon. At the end of the spit where the lagoon opened into the sea the coast was littered with the flotsam and jetsam of the obliterated community, as it slowly made its way out of the lagoon and into the sea. Although some of it had been washed so far inland across the 5km (3 miles) wide lagoon and into the fringing mangroves that it would never find its way out.

And what of the people who had lived there? I met a man one day who looked shell shocked to say the least. He was an aid worker who had personally buried over 30 bodies, and in the baking tropical heat the task got less and less pleasant with every passing day. As we traveled by canoe over the lagoon we heard that every day people were pulling bloated bodies out of the water. They had been washed into the lagoon and covered with tsunami sand only for the gas build up in the putrefying bodies causing them to float up to the surface days and even weeks later. The more time that went by the harder it was to remove them from the lagoon because in pulling them out and into a boat, the flesh just slid off the bones. The simplest expedient was to use a net in a bizarre daily fishing expedition. The village was so remote that even when we were there some four weeks after the event there were still bodies, well skeletons by then, draped over trees in the mangroves or simply covered by piles of debris to be removed later. Though some bodies had been removed already. We heard of sharks coming in at high tide, sliding up the beach, grabbing the nearest bodies and then simply allowing gravity to slide them back out to sea, dragging the bodies with them.

These are not experiences I wish on anybody, but this is the reality of post-tsunami carnage.

Of more concern is the desire to return to the spit to live. While on site we found clear evidence of past lagoon subsidence, past tsunami inundations, and repeated tsunami hell over the decades. Abundant ancient stumps of drowned trees could be seen through the shallow waters of the lagoon, submerged former shorelines were found, and we could see evidence for at least two past events from around 1934/1935 and 1907. In fact, there have been at least seven tsunamis along the northern cost of PNG since 1907 that we know of and given the time that has passed since the most recent one it is very likely that next one is due any day now.

This part of the world gets a lot of earthquakes, and the hills behind Sissano Lagoon, the Barida Range, have been lifted up about 52 m (170 ft) by earthquake activity over the past 6000–7000 years. This seems like a staggering amount of uplift, but when averaged out that comes to about 1 mm (0.04 inches) each year. We know all of this because of the Aitape Skull. It was found in the left bank of Paniri Creek, in the Barida Hills, about 12 km (7 miles) immediately inland from Sissano Lagoon at an elevation of around 52 m (170 ft). It was in sediments that had been intertidal, probably an old lagoon, showing that the land must have gone up 52 m from the then sea level some 6000–7000 years ago.

The Aitape Skull was originally discovered during a geological survey in 1929. The original report simply states "In addition to the foregoing, a carbonised coconut was found, together with a fragment of human skull. A search for teeth or other relics was conducted without success." The skull was discovered by a team member, Paul Samuel Hossfeld, during their brief four-hour stay in the area, but since the team was working for an oil company they were somewhat focused on that side of the work. After their trip the skull was given to the South Australian Museum.

Everything went quiet until 1941 when the cardboard storage box was dusted off and the skull was finally examined by a man called Frank Fenner. He considered that the skull was probably that of a female, around 45 years old, and that it was the first evidence from PNG of human remains from the Pleistocene. In other words, it was thought to be more than 1 million years old! This was big news, very big news. In archaeological terms it suggested that our ancestors had moved out of Africa and as far east as PNG extremely early on in human history. From here they could have got across to

Australia far earlier than anyone had ever thought. This was a bit like finding Neanderthals in Hawaii!

Hossfeld finally picked up the reins again and had another look into this in the early 1960s. He even undertook another trip to the area to try to find more of the skeleton and any other material that could be dated. Unfortunately, the creek was in flood at the time and so all they managed to get hold of were some shells, more coconut, and pieces of wood that were all encased in a small lens of sand within this intertidal sandy-mud deposit. The basic interpretation was that the environment represented by what they found was a coastal mangrove with a sand layer in it.

But Hossfeld now had more dateable material to go along with the skull and used the then exciting newfangled, and all the rage, radiocarbon technique to get some dates. Oh dear, the skull was only about 5000 years old. Needless to say, this rapidly assigned the skull to the obscurity of its dusty cardboard box again because it fitted perfectly well with what was known about human use of the area at that time. No archaeological smoking gun, just another part of the well-trodden human story of the area.

Sigh . . . how dull.

But no, the old cardboard storage box was dusted off once again in 1996 and further analysis suggested the skull belonged to a he, not a she. The skull was continuing to throw up new bits of information. But wait, there's more, and here we come to the fun bit, well I think so anyway. After our work following the 1998 tsunami my colleagues and I wrote a paper about it all and in the process we found all of the early papers about the Aitape Skull. The descriptions of the environment in which it was found piqued our interest and we compared it with what we found at Sissano Lagoon following the 1998 tsunami. We even said "the sequence (where the Aitape Skull was found) is strikingly similar to the . . . sedimentary sequence, and state of many of the human remains in Sissano Lagoon." At that point we had just a sneaking suspicion that maybe, just maybe, the Aitape Skull and the place it was found could be interesting to look at.

Nothing happened.

We had jobs to do and no research funding anyway (that old refrain), so it just faded into the background . . . again. It was purely by chance that the skull came back into my brain cell. I was working with archaeologists from the United States who knew the area well. After trying for about 10 years, we somehow managed to get US funding to go and look at the early human occupation of the area and to try to understand how they interacted with their environment—cool. Such things never run smoothly though, after all the Aitape Skull itself has had a checkered career.

To get to do research in PNG you need a research permit, you can't just sneak in on a tourist visa and do the work. Well, you could if it wasn't for the fact that hardly anyone wants to go there as a tourist, and so it might look a little strange turning up with a bunch of geology field gear and saying that you are there to sit on a beach somewhere.

So we all applied for a research visa.

The US guys got theirs, it took a few months but they came through. I dutifully filled out all the forms, took my passport in to the PNG Consulate in Sydney, they took my passport (gulp) and kept it for a few weeks while they processed the application. A month later an email arrived (only a week before I was due to leave) saying it was ready. Off I rushed to the consulate again, there was my passport (phew) and the research permit firmly stuck inside (phew). Just as I was leaving, I glanced at the date on the permit— it had expired the day before! No, it couldn't be changed, a new application would be needed and there was no time. My colleagues went without me, they visited Paniri Creek, collected more samples, and I festered and mumbled and grumbled in my office. When they returned we analyzed the samples collected, reviewed the detailed site descriptions of my colleagues, and compared these with the interpretations of Hossfeld and his colleagues.

Wow—the sediments, the geochemistry, the shells, wood, microfossils, and the structure of the deposit all cried out one thing—the skull was embedded in a tsunami deposit! This was, and still is for now, the oldest known tsunami victim in the world.

You would think that was the end of it, but two things came out of this. First, we wrote up all of our findings and published them. Well done. In this process we thought it would be useful to have another look at the skull, after all you never know, we might see something new. There was only one problem though, no one was allowed access to the skull anymore because it was in the process of repatriation to PNG. The dusty cardboard box was on the move again. Second, I was rather pleased to see the paper come out and one day as you do sometimes, I was excitedly talking about it to my partner, who incidentally knows a lot about tsunamis . . . and archaeology. I was explaining (probably more like mansplaining) how the skull was one of the archaeological criteria we could use to help identify prehistoric tsunamis which was pretty cool, but that the list of criteria, which had grown quite rapidly in the early days, was somewhat stagnant now. I really wanted to try and find new ways of identifying prehistoric tsunamis. Yes, we had started to wander into DNA found in sediments and other esoteric areas of research, but really it was all a bit ho hum to be honest—just another bit of stuff in the sediment to analyze. I felt that the discipline was starting to look a bit inward, just more and more microscopic analysis of sediments and not much else. To me the real fun was in what the geologists would think of as the peripheral stuff, something like archaeology for example. After all, tsunamis kill lots of living things and rampage through coastal communities. It seemed to me that if there was anything exciting and new to do it would somehow relate to this side of the equation.

She said, "What about mass burials?"

It was as simple as that, or as complicated as that.

I happened to be talking to probably the only person in the world who could have seen that, who grasped the importance of what every other tsunami researcher had missed. We worry about tsunamis because they kill us, a lot of us if it is a bad one. In the 2004 Indian Ocean event there were 14 mass graves in Banda Aceh alone, the largest of which contained 60,000–70,000 people. In the immediate aftermath of such an event there are usually attempts to bury individuals in culturally appropriate manners, but mass burial soon becomes the norm as the sheer number of victims simply overloads the system. What about in prehistory? OK, so the coastal communities were not as large as they are today but there is absolutely no reason to suspect that if such an event occurred in the past there would be no deaths. Also, in prehistory there was no international aid, no medical SWAT teams to help clear up the mess. Mass burials would have been inevitable, so where were they?

8

Strand 7

Piled Higher and Deeper

During the site survey, scattered human bones were seen in various places, indicating isolated burials, but at five sites there was definite evidence of a burial ground in the form of a low mound with considerably more concentrated evidence of human remains. Surface evidence on these mounds included not only scatters of fragmented human bones, but also shell money beads, pieces of Trochus arm rings, and worked Nautilus shell.

B. Leach and J. Davidson, The Archaeology of Taumako

"What about mass burials?"

It was a light bulb moment. For me anyway.

It is always nice to be brought down to earth from any particular ivory tower you might be sitting in, knowingly or unknowingly. My partner had cut to the chase. For several years I had been getting more and more interested in the human side of tsunamis through both tsunami education and a better understanding of the archaeological aspects of the topic. In just four words she managed to bring both geology and archaeology together. After all, a mass burial at or near the coast increases the likelihood that the deaths were probably related to some nasty event—maybe a cyclone/hurricane/typhoon or a tsunami or even a shipwreck.

I'm getting ahead of myself a bit here. I am committing the cardinal sin that Chamberlin alluded to, "the habit of some to hastily conjure up an explanation for every new phenomenon that presents itself." The point is that all possible interpretations need to be considered.

I caught myself just in time there.

In Search of Ancient Tsunamis. James Goff, Oxford University Press. © Oxford University Press 2023.
DOI: 10.1093/oso/9780197675984.003.0008

So, in order to avoid bias it was important to recognize that studies of mass burial sites involve the great fieldwork skills of archaeologists. They have dug many mass burial sites, some at the coast and many not. Over time a narrative has developed that invariably places the archaeological interpretation of mass burials into one of several easily recognizable categories such as ritual sacrifice, war, famine, plague, or possibly an occasional natural disaster such as an earthquake.

On occasion it is difficult to crowbar the archaeological narrative into any specific category, but nevertheless the data collected during fieldwork and in subsequent laboratory analysis allows commendable speculation about diverse topics such as variations in burial position, mortuary practices, status of the people buried, grave goods, diet, general living conditions, community structure, and so on. Once considered in so much detail the actual raison d'être can easily be ignored or forgotten, or merely speculated upon in passing. This is entirely understandable because the focus is on what the skeletons and burials say about the culture, but often focusing on the detail misses the main event—why was there a mass burial in the first place?

Interestingly, there are not as many prehistoric coastal mass burials as we would expect. To be honest there are probably two main reasons for this. First, they simply haven't been found yet (or looked for). Second, sea level rise, coastal erosion, increased storminess, and a bevy of ongoing coastal processes have undoubtedly wiped some of them off the planet.

Of the mass burial sites that have been discovered, some have been prehistoric, others are in the early historical records, and many sadly, given what we have been through over the past few hundred years or so, are more recent. The latter are generally war graves or conflict related, but epidemics and natural disasters have accounted for some. In many cases, identifying the cause of a mass burial is pretty easy—skeletal evidence for decapitation, blunt force trauma, and weapon injuries all point inevitably to violent deaths associated with warfare. Here though we are on the search for possible prehistoric coastal mass burials related to tsunamis.

First off it is worth noting that I was entering into an area that is most definitely within the archaeological domain, and I am a geologist. There are established narratives, there are accepted interpretations for coastal mass burials, but it is also worth noting that there has never ever been a single one assigned to deaths related to any form of high energy marine inundation. I had a feeling that by the end of all of this I would have rattled a few people's cages, even ruffled a few feathers. But it is a topic well worth exploring and so should not be shied away from (Plate 8.1).

It was ever so slightly disingenuous of me to suggest that there has never ever been a single coastal mass burial assigned to any form of high energy marine inundation, but as you will see, it is probably a fair statement. As an example of how mass burial interpretations are not as easy as they seem, let us first look at a coastal site in Greece to get an idea of how the land lies.

Archaeologists working on ancient Greek sites have a slight variation on a theme for their interpretations of most of the mass burials they have found. Most are considered to be casualties of warfare, or victims of an epidemic or famine, or prisoners of war or convicts. In one instance though there has been a slightly innovative interpretation—a mass burial of 115 people found in the small town of Pydna in coastal Macedonia, in northeast Greece. Pydna sits pretty much at the head of the long funnel-shaped bay known as the Thermaic Gulf, which points southeast toward western Turkey and the Aegean Sea. It is a lovely part of the world.

Radiocarbon dating (that old trusty technique again) indicates that this burial probably took place in the second half of the 4th century BC (300–400 BC) when Pydna was an important Macedonian city. The bodies appear to have been tossed into a deep rock-cut shaft and forgotten about. Archaeologists recognized that these people were low status because they had been in poor health, they had a lower than average age at death, their skeletons indicated lots of heavy physical stress even to the point that some were in iron shackles. An entirely reasonable inference therefore was that this might be evidence for a mass burial related to ancient slavery.

OK—I feel that we need a bit more detail before we can really get on the track of a tsunami. We need a bit more context. First of all, there were at least 42 adults, 16 sub-adults, children and juveniles of both sexes. Not your average slave population but indeed it represented a balanced population demographic of both sexes. Agreed, they were probably all low status, some may have been slaves (shackled) and some not, but this may simply have been a good indication of the general population at the time. A further wrinkle comes along when one considers the landscape archaeology of the region.

Just on the Aegean Sea side of the Thermaic Gulf lies the western end of an earthquake-prone patch of seafloor known as the North Aegean Trough. The trough lies at right angles to the gulf. In other words, if there was a decent underwater earthquake, the way the fault moves means that it could fire a tsunami directly into the gulf. Couple that with the fact that perhaps the worst place to be when a tsunami approaches the coast is at the landward end of a funnel-shaped bay or gulf. As a tsunami moves into the gulf from the Aegean Sea its energy gets focused or squeezed by the land on either side— more energy in less space—and this gets more and more pronounced as the wave moves up the bay until it can release all of that energy as it inundates the low-lying land at the end of the gulf.

Right, so not necessarily a good place to put a city. We have the ben- efit of hindsight today though and can wisely pontificate with such scien- tific pronouncements. But at the time no one knew about faults and their tsunamigenic potential, so they can be forgiven for putting themselves in harm's way. Unfortunately for them, the earliest known tsunami-generating earthquake here occurred in 330 BC (that's the 4th century BC) near Limnos Island, a mere 220 km (140 miles) away by tsunami. Having said that, there have been no reports of any tsunami deposits being found at Pydna even though it would have been in the direct line of fire. One suspects that this might simply be because no one has looked, but we do know that the tsunami struck the Turkish coast some 180 km (110 miles) to the northeast, a direc- tion that would have received far less of the tsunami's energy.

Needless to say, archaeologists have done a great job at studying the burial context and skeletal remains, and they have even come up with a new pos- sible type of mass burial—just throwing a whole bunch of slaves into a pit at the same time. That does rather beg the question of why throw in a bunch of children—OK, they could be slaves, but maybe not? And, while all were of low status, none seemed to have been particularly incapacitated physi- cally. To my mind, and I admit that I have neither visited the site nor seen the skeletal remains first-hand, this "new" slave burial interpretation seems to simply be a variation on a theme within the existing paradigm of archaeo- logical thinking with respect to mass burials. It is well known that Pydna sits in an earthquake-prone region, but there has been no consideration of the possibility that this could be a mass burial related to something such as the 330 BC earthquake and tsunami. That is unfortunate because the date of that event matches rather nicely with the date of the mass burial.

I should not be too harsh. To start with, the tsunami paradigm needs to get greater penetration into the world of archaeology, and heck—I miss many things as well. Not only that, but it is also to Greek archaeologists that we must look for the single example of a mass burial that might possibly be re- lated to tsunami inundation . . . possibly.

It was the Greek archaeologist Charalambos Kritzas who suggested the idea. He worked at Argos in the eastern part of the Peloponnese Peninsula in southern Greece. It sits at the head of the Argolic Gulf, which faces southeast directly toward the island of Santorini 270 km (170 miles) away (yup, that's the one that blew up with the tsunami it generated being responsible for so much carnage in the Mediterranean, leading for example to the decline and fall of the Minoan culture on Crete. You can read more about this in my book with Walter Dudley—*Tsunami. The World's Greatest Waves*). Kritzas found a well containing 20 human skeletons of various ages and sexes, ten males, six females, and three infants were identified. There were also a lot of animal skeletons (a horse, cows, pigs, sheep, goats, and a dog). This strange mass burial was dated by assorted methods, such as pottery types and other artefacts, to around the Late Helladic period around 1250–1300 BC.

Kritzas thought that the inclusion of animals in a human mass burial was unusual, and as such required an unusual explanation. He tentatively concluded that these were all flood victims. OK, fair enough, this does not actually mean a tsunami was the cause but let me argue the case. He did not directly suggest a tsunami as an explanation for the burials, but he did make an important reference to the writings of a Greek traveler and geographer of the 2nd century AD, one Pausanias, who in his travels around Greece, wrote about the Sanctuary of Poseidon, "for they say that Poseidon inundated the greater part of the country because Inachus and his assessors decided that the land belonged to Hera and not to him. Now it was Hera who induced Poseidon to send the sea back, but the Argives made a sanctuary to Poseidon Prosclystius (flooder) at the spot where the tide ebbed."

To untangle that just a little. In Greek mythology Inachus was the first king of Argos, and a resident of Argos is called an Argive. Hera was the goddess of women, marriage, family, and childbirth and was worshipped as Argive Hera at her sanctuary, which stood between Argos and Mycenae (11 km, 7 miles, north of Argos). I have been unable to find out where exactly the site of the sanctuary to Poseidon Prosclystius is located, but I assume that it is probably somewhere between Hera's sanctuary and the sea.

This is relevant to Kritzas because he suggests that this story may refer to the flooding of the plain of Argos by the tsunami associated with the eruption of Santorini (Thera) around 1350 BC.

Bang—there you go. This is the only possible, tentative, interpretation that links any coastal mass burial site in the world with a tsunami. So I therefore argue that until very recently when we stepped in with our research, not a single prehistoric coastal mass burial site had ever been assigned to deaths related to a tsunami. In historic times there have obviously been quite a few mass burials following recent events. But the further you go back in time the less documentation there is, and you eventually drift into a situation where while a site may technically be an historical mass burial, in reality there is absolutely nothing written about them so they might as well be prehistoric.

Many countries around the world have a long written history, and so we can benefit from those in helping us to better understand what might have gone on in places where there is a much shorter historic record. I mentioned earlier that Japan has a deep written history with references to tsunamis as early as AD 684, but even there these older writings are moderately few and far between. They are not like today's media that capture almost anything going on anywhere in the world if you care to look for it. These Japanese writings are snippets of history, recording an event for example, but not providing a lot of detail.

If we come forward in time about 600 years, the AD 1293 Kamakura earthquake and tsunami were documented in Japan. While this is still just at the cusp of a decent Japanese written history, the details are sufficient to add immense value to what would otherwise just have been a full-on archaeological investigation, narratives and all.

Written accounts of this event are found in two books, the *Kamakura Oh Nikki* and *Azuma Kagami* and state that some 20,000–30,000 people died, most in the tsunami, with the bodies of humans, cows, horses, fish, dolphins, and whales filling the roads. This now makes the interpretation of the Argos mass burial sound more reasonable, after all this is a written record and you don't just have to believe what I say. The entire Kamakura region, just south of Tokyo, was cut off from the rest of Japan, and Kamakura city was in ruins. It therefore comes as no surprise to learn that many of the dead were not buried for some time. But where was the mass burial? The historical accounts did not provide an x to mark the spot, they are simply documents providing some details about the event.

Dammit.

But the rapacious growth of Japan's urban infrastructure demands new buildings, new roads, cars and car parks, railway lines and trains. It was only a matter of time before a mass burial was found.

Yuigahama beach provides a wonderful waterfront for Kamakura city, and it also provided one of the only areas to build more car parking space . . . underground. And there it was—3108 human skeletons were discovered, many with dog bite marks showing that they had been lying on the ground for quite some time with body parts scattered by dogs and other scavengers before they could be buried. Animal skeletons were there as well, just as the historical accounts had said. They too had been victims of the tsunami, with the dolphin and whale skeletons giving some sense of the power of this event (and also showing that the tsunami interpretation in Chapter 1 for the slightly bizarre whale skeleton find on a clifftop in New Zealand is not as whacky as you might have thought).

There could have been a wrinkle though because in AD 1333, just 40 years after the earthquake and tsunami, there had been a great battle when Nitta Yoshida attacked Kamakura. Nitta Yoshida was 35 years old when he was sent on imperial order to defeat the Hōjō clan who held Kamakura. As he attacked, Nitta Yoshida (AD 1301–1338) found himself at an insurmountable impasse. He could not directly assault the fortress there because it had almost vertiginous slopes and Hōjō's fleet was spread out on the ocean behind him blocking him in. It is said that Yoshida placed himself by the shore and looked out toward the ocean, praying to the dragon god Ryujin, the all-powerful ruler of the seas. He then threw his sword into the water and the dragon god parted the sea providing him an easier approach to the Hōjō stronghold. The more prosaic explanation is that he waited until low tide, but let's not ruin a good story.

Where was the mass grave from this attack? Was it the one that had been found when they were digging the hole for the carpark? There were some good power struggles going on in the region during this time and so it may well be a candidate. BUT, and this is a very important but, only about 1% of all of the skeletons from the Yuigahama beach mass burial had sword cut marks PLUS the general age distribution of the skeletons mirrored the overall demographics of the population at that time, not those of a bunch of warriors cut down during the battle. This was a burial of tsunami, not warfare, victims. Are we convinced though? Do we think that there is enough evidence here to support this hypothesis? For most historically documented tsunamis there is little reason to mention the association of the tsunami deposit with the mass burials, look at the 2004 Indian Ocean tsunami in Banda Aceh. We know there were mass burials, who cares about the tsunami deposit?

Kamakura is slightly different. There are some complicating factors such as wars and war graves, but in this case though we are safe yet again, because there is a considerable body of geological evidence. This raises an important point, if we are trying to say that a prehistoric (not historical) coastal mass burial was in fact related to a tsunami, then at the very least we should look for geological evidence for that event and date any findings. This is an important step in establishing a possible link between the two.

I visited Kamakura and the car park, the beach, and the fortress. It was in many ways just another Japanese coastal city, but on the other hand it was not, it had a tangible link with a frightening disaster that had occurred nearly 800 years ago. If it could happen in Japan, it could happen anywhere around the Pacific at the very least and let us not forget Greece, and actually almost anywhere with a coast. This Japanese mass burial was most definitely the result of a tsunami—we had all the data we needed to prove it. What about elsewhere?

The search was on. The game was afoot.

While I have studied tsunamis all around the world, my "patch" if you can call it that would probably be the Asia-Pacific region, about half the world, but who's counting. I love the Pacific in particular, there are so many islands, so many different and yet similar cultures, so many tsunami sources, so many oral recordings, so much geology, and lots of archaeology. If there was anywhere in the world that should have some evidence somewhere, then it would have to be in the Pacific Islands. It was also incredibly likely that some tsunami-related mass burials had already been discovered but not recognized as such. After all, most Pacific archaeologists are of the opinion that natural hazards such as tsunamis are essentially pretty irrelevant when compared with the effects of other impacts such as climate change and human modification of the landscape. A fair enough point of view when you have not really been exposed to the consequences of major tsunami inundation.

Hmmmm—OK, so my thoughts that tsunamis might have been responsible for some of the mass graves that have been found may go against the grain in some cases. After all, field archaeologists have worked for years at such sites to draw out the minutiae of details related to the burials. There was nothing for it but to jump in with both feet and start trawling through archaeological reports and papers.

An impressive amount of archaeological data has been produced from research in the Pacific. It was an insurmountable task to go through it all and so I focused on three keys things. First, I would search for evidence in an area where I knew there had been big events in the past, the southwest Pacific. Second, I would look for any mention of burials, single or otherwise. Finding archaeological reports about a mass burial was the easiest option, but finding evidence for single coastal burials might also lead to some interesting discoveries. Third, my search would be limited to dates around the timing of the two most significant earthquakes and tsunamis that have occurred in the southwest Pacific in prehistory, AD 1450–1500 and 2800–3000 years ago.

The archaeological quagmire that is the southwest Pacific opened up before me. After considerably more than a century of research in the region, I probably had too much data, a case of not being able to see the wood/forest for the trees, as the old saying goes. Let us start with old school archaeological interpretations and then look at a "new" school example from more recent efforts. In reality just as in geology, what this seems to show is that the more recent work is simply "more of the same," just in more detail. We scientists have more toys to use, and we use them.

A case in point, and completely unrelated to mass burials, or at least it might be, is some recent work I have been involved in related to Easter Island/*Rapa Nui*. Sadly, no, I did not visit Easter Island/*Rapa Nui* although it may be on the cards in the near future. A group of us—myself and geologists and other "geo" types from Spain and Chile—put together a review of all of the geological processes that influenced the human occupation of Easter Island/*Rapa Nui* through time. This was really the equivalent of a landscape archaeology review that, amazingly, had never really been done before. In other words, all of the current thinking about what humans did and why on the island have never been properly placed within the overarching framework of the environment in which they lived/live. Ugh.

A word of caution here, Easter island/*Rapa Nui* is essentially famous for a remarkable cultural collapse that has been written about, discussed, argued about, covered in numerous documentaries, and for which we will never know the full story, but it is nice to try and have as many ideas as possible thrown in to the mix because one of them may end up having more traction than the others and finally lead us to some conclusions.

The creation of *Moai* (those massive legless bodies that can be found standing, if that is the right term for something without legs, in so many places around island), the rapid loss of the native forests, the warfare, etc., etc. There has been so much amazing work done on Easter Island and the archaeological output is incredibly detailed, but the physical environment has been less well covered. I will not dwell on all of the work we did, suffice it to say that the variability in volcanic rocks determined what was mined and where and also how good the soil was, the availability of water sources determined decisions on land use, climate change also played its part, and so on.

Amazingly, until one of my fellow co-authors put together a paper in 2018 no one had really considered tsunamis much at all, other than as an interesting aside, a minor blip in the big picture of the island's human history.

I should say now that I entered into my bit of this work with no major preconditioning or points to make. However, I was aware that there was quite a lot of historical evidence for devastation caused by the 1960 Chile tsunami on the island's eastern coastline. This is an event we have come across before and will also see more of later. On the island, waves over 10 m (33 ft) high destroyed the famous *Ahu Tongariki* and carried its debris over 1 km (0.6 miles) inland. *Ahu Tongariki* is the largest *ahu* or stone platform on Easter Island/*Rapa Nui*. *Ahu* are ceremonial/sacred sites and in the past had been used for burials. These were not "mass" burials in the sense that a whole bunch of bodies were all buried at the same time, but there were multiple burials, it is just that the timings of these are unknown. Most *Ahu* are found on the coast with the odd exception on higher ground. One thing that is notable about *Ahu Tongariki* is that the *Moai* face away from the ocean, facing sunset during the winter solstice.

It was not the first time that the *Moai* that sit on top of *Ahu Tongariki*, and the *Ahu* itself, were destroyed (Figure 8.1). The generally accepted interpretation is that their previous destruction took place during cultural unrest sometime after AD 1500. In 1960, the *Ahu* on which the *Moai* had stood was completely destroyed and several *Moai* weighing up to 50 tons were carried more than 150 m (490 ft) inland. As the tsunami moved inland it littered the area with boulders, dead sheep, ripped-up seaweed, fish, and human bones from the burials that were associated with the *Ahu*. Importantly, the site must have been restored in prehistory so that it could be destroyed again in 1960.

Figure 8.1 *Ahu Tongariki*, Easter Island: panorama looking northwest taken in 1914/1915, showing what was then, and had been for a long time, the "intact" *Ahu* with only the lower pedestals of the *Moai*. The present day reconstructed *Ahu* was completed in the 1990s following its complete destruction during the 1960 tsunami. (Photo source: British Museum, Public Domain)

Amazingly, once again, *Ahu Tongariki* was almost completely restored over a five-year period in the 1990s through the incredible efforts of a multidisciplinary team headed by archaeologists.

History repeats itself?

It is fascinating to learn that *Ahu Tongariki* was considered so iconic (I could cynically say that this is probably because it is one of the island's most important tourist sites) that it was reconstructed again in the 1990s. In a way this provides something of a clue for us when we consider that it must also have been an incredibly iconic site in prehistory too. Does this destruction/rebuilding cycle offer us possible clues about the devastating effects of one or more paleotsunamis? What would an earlier event of similar or greater magnitude to the 1960 event, and from a similar source, have done to the then coastal population, monumental architecture, and resources in general? The precursor to the Magnitude 9.5 1960 Chilean earthquake and the tsunami that destroyed *Ahu Tongariki* occurred in AD 1575. This may have even been even larger than the 1960 earthquake but there is so little historical data from the early colonizers in Chile that it is hard to be definitive about that, suffice it to say that it was big.

There are some intriguing confluences of indirect evidence for the possible effects of the AD 1575 tsunami (which just sneaks in as an historical event on mainland Chile) on Easter Island/*Rapa Nui* (where it was a paleotsunami). First, it is generally recognized that the initial destruction of *Ahu Tongariki* took place during cultural unrest "sometime after" (key weasel words) AD 1500. I dug back as far as I could to see where this date came from—it seems to now be one of those dates that all data "fit," but actually is pretty hard to tie down. A paper written in 1973 seems to suggest that this unrest was probably sometime before about AD 1550–1650, although the errors on these dates are substantial, so who knows. The point is that that the AD 1575 tsunami fits the bill here, and the bill in this case is cultural unrest, cultural change, population decline, inland migration of the main settlements, and let us not forget, the destruction of the *Ahu*. All this points toward the possible devastating effects of the AD 1575 tsunami on Ahu *Tongariki*. I would just like to add one minor extra snippet of information here. Some scientists have also pointed out that some of the *Ahu* reconstructed "sometime after AD 1500" were rebuilt out of their older selves together with an array of marine material. This in itself may well indicate that the local people used resources immediately available to them, found among the *Ahu* rubble, material that was left behind by paleotsunami inundation. This is very neat, and it makes me wonder about the use of *Ahu* for burials. *Ahu* are essentially coastal burial sites OR where they are not at the coast they are on the highest ground available—away from the potential tsunami inundation zone? Could these have been mass burial sites related to earlier tsunami inundation? I think this is probably one speculation too far!

As tangents go, the Easter Island/*Rapa Nui* example did at least end up having some relevance to mass burials, but now back to other parts of the Pacific.

There is a coastal mass burial site on the island of Taumako, in the southeastern Solomon Islands about 400 km (250 miles) northwest of Anuta, which already had an honorable mention in the last chapter.

I have literally only just realized the relevance of that statement with regard to the age of this mass burial! All of this stuff around the Solomon Islands dates to around the 15th century and one of the big Tonga Trench earthquakes and tsunamis. This mass burial I am about to talk about is dated to the 15th century, which is also around the time people escaped the Tongan invasion on Wallis following the event—gosh, it is amazing what writing a book can do in helping join up even more dots, amazing.

OK, armed with this new link to even more evidence for the 15th century event, let's have a look at the Taumako mass burial. Archaeological fieldwork was carried out here back in the late 1970s, but the subsequent analysis and final report writing meant that we didn't get to learn much about what they found until 30 years later. These things take time. The burial site contained somewhere between 201 and 226 individuals, depending upon how you figured out the skeletal remains—the comingling of bones made it hard to be precise. The bodies had simply been heaped up on top of each other with a thin sand layer and some coral covering them in some places. No grave digging seems to have taken place. This was not part of any traditional Polynesian burial practice (I should point out that while the Solomon Islands are predominantly Melanesian, Taumako is a Polynesian outlier).

The general sense of disorganization in this mass burial caused something of an archaeological narrative dilemma. It was observed that the burials were so close together that if they were not actually all buried at the same time then the interval between the burials was so short that the people burying new bodies would have known those from previous burials that were effectively lying on the surface—yuck. Hmmm—not exactly traditional burial practices to simply bung your parents on top of say your grandparents without any ceremony and just leave them there to rot. Somehow, and I found it fascinating that this became a key point to the interpretation, by accepting that this was actually what they considered happened the archaeologists went in a different direction and identified what they saw as fourteen potential burial layers within the mound.

Six radiocarbon dates were done and if you merged them all together the pooled age ranges for the mass burial were between either AD 1220–1745 or AD 1275–1655 (depending upon how you played with the statistics), a difference of 525 and 380 years respectively. Well to start with I wouldn't have merged all the dates, each one needs to be examined separately, but hey (and statistically, it is only the maximum potential age range for the site that should be considered, so the shorter age range of 380 years was largely irrelevant). Following these calculations, the maximum (525 years) and minimum (380 years) age ranges were added up and then divided by two to get a midpoint. As a result, the mass burial apparently represented 450 years of burials (380 + 525 divided by two). Hopefully this all makes sense—it is not logical or good science, but I hope you can see how it was done.

Of course, the flip side of all this fancy footwork with a bunch of dates is that when you look at the age ranges of each of the individual radiocarbon dates there is absolutely nothing between them at all because they all overlap for some dates within the AD 1220–1745 range and so as such they can quite easily all relate to a single event or day. Playing with radiocarbon dates to try and fit them into a single narrative you want to tell is really not helpful at all.

It is therefore slightly worrying that while acknowledging that there was no way of knowing what the actual time gap between any one of the "layers" might be, it was decided to infer an average of about 32 years between each layer (the mid-point of 450 years divided by 14 layers). This then allowed the researchers to head off down that particular narrative route of multiple layers and not acknowledge or consider other possibilities (I feel Chamberlin and his Multiple Hypotheses turning in his grave). To be honest, to my mind the final report contained a lot of data that pointed to the very real possibility that what was being investigated here was a paleotsunami related coastal mass burial, not some long-term burial site where the dead were just heaped on top of their recently deceased relatives. Let's examine the data.

- First, the method of burial was consistent with rapid burials. The site is underlain by coral, not a great spot for a burial ground. Bodies were all piled up on each other with no traditional burial practices performed.

- Second, the age distribution of the skeletons represented that of the population at the time, not just old people—old, young, male, female.
- Third, *Kilikili*—surface scatters of coral gravels were found in assorted places around the island (I have already talked about tsunami-lain gravels scattered over sand dunes up to 32 m, 105 ft, high in New Zealand).
- Fourth, arrays of scattered human bones and other concentrated mounds of human remains (mass burials) were also found in other places around the island . . . undated though.
- Fifth, on an island just to the north a sand and gravel deposit was found up to 50 m (165 ft) inland, dating to between AD 1299 and 1470. Hmmmmm. OK, so that is a decent age range for a radiocarbon date but, just like that for the mass burial site, it covers the time period for the mid-15th century Tonga Trench paleotsunami.
- Sixth, oh yes and another important gem, there are several oral histories about large waves affecting many of the islands in the area in prehistory.

In addition to all of these points, the island of Taumako is in the line of fire from several different underwater earthquake sources, including the Tonga Trench and also the Solomon Trench. But also, given the amount of groundshaking associated with really large earthquakes, there could have been underwater landslides that generated local tsunamis, and there is always the possibility of a volcano popping off in the neighborhood. We are replete with tsunami sources. Having said that though, we are also replete with tropical cyclones in this neck of the woods too, and there is the possibility that this coastal mass burial site could be cyclone related. I am inclined to not go down the cyclone route because there are several things against that suggestion, most notably the coastal setting and the fact that, unlike a tsunami, the local community would have known that a cyclone was on its way. It is the unexpected catastrophe that leads to such mass burials.

On a final note, while there are no tsunami oral histories specifically related to this site, the place is called Namu and this can be translated from Polynesian to mean "bad smell." The name Namu may represent the last remaining cultural evidence of a hastily dug mass burial site. It also suggests that this was not a regular cemetery representing 450 years' worth of really poor burials, imagine the smell.

OK, so I have a nice warm feeling that this example is pretty much guaranteed to have been a coastal mass burial site related to a paleotsunami.

Yes, this is my interpretation and I put my hand up and own that—the dating fits (it is a pretty wide date range, so that is not difficult), the location fits, and the nature of the burials fit. The evidence from the geology fits too, but the icing on the cake would be to go there armed with our new glasses and give the place a thorough geological examination. Will that happen? Probably not in my lifetime.

The earlier Tonga Trench event that I was hopeful might throw something up was from 2800–3000 years ago. This was an earthquake and tsunami that seemed to be the equal of its younger cousin. However, the further back in time you go the harder it is to find skeletal material. Obviously it is by no means impossible, but such a lot of things have happened over thousands of years compared to hundreds of years that evidence gets lost. Landslides can take things away, the sea can erode the coast, the tropical climate can cause rapid decay, people can disturb the land, and so on. The list is almost endless. So imagine my surprise when I read about an incredibly important coastal cemetery/mass burial site in Vanuatu. From an archaeological perspective it is exceptionally important because it contains the earliest and most extensive collection of burials associated with early human settlement (Lapita—Proto-Polynesian) of the Pacific Islands, and dates to around 2800–3000 years ago. Bingo! However, and this is an important however, I have an internal warning system that starts ringing bells when I immediately jump to a conclusion and go something like, bingo!

It is important not to get stuck in a one-track argument and it is vital to consider all possibilities (good old Chamberlin again). On the flip side, though, when I look at most archaeological reports from work done in the southwest Pacific there is just that, a linear argument without considering all the options. So as opposed to going through the process of repeating all that has already been said, in this case I basically did what I had done before and took all of the reported data and analyzed it to see if it could be interpreted in any other way, after all, it had already been interpreted one way.

First of all, this was a modern, 21st century discovery, which meant that there was a lot more fancy footwork involved in the archaeology which is great. Let's tick off some of the key points:

- The site is right next door to what was the beach at the time with shallow graves dug into sediment that fills gaps in an old reef—not a great place for burials.

- There were 91 individuals of which about 80% were adults (of both sexes) and the remainder infants, a good representation of the population.
- A lot of radiocarbon dating was performed on an assortment of things, which was great, and then a bunch of statistical wizardry was done to try and tease out any patterns from the dates. In essence this is a way of simply trying to use the statistics to infer where some dates are more likely to represent a certain date range than others. In other words, can we find differences? Another way of saying this is—these dates can't all relate to a single event can they, there must be more to it than that? Unfortunately, once again, this starts to bias thinking away from the possibility that these burials may relate to a single event. Not surprisingly therefore the cemetery was considered to have been in active use for over 150–250 years or so. HOWEVER, and this is good, in this case it was noted that what might have been considered dates related to different phases of use could actually entirely overlap and so may equally represent burials related to a single event, or at the very least with the majority of burials occurring at one time. Alright, so it could be a mass burial, but that is where that line of argument stopped.

At this site things are more complex than in the Solomon Islands example and a case for it being tsunami related is less clear, but there are some pointers which need to be considered. First, the old reef area that was by the sea at the time would appear to have had to accommodate both a settlement as well as the burial ground, and there wasn't much room for that. This is unexpected in day-to-day life in Polynesia, a massive burial site literally on the doorstep. But if it was a mass burial related to deaths from a tsunami then this is a pragmatic response—bury everybody at the place where there are the largest number of victims, the destroyed settlement, so they were actually not both sharing the site at the same time. There was not a lot of ritual burial such as body manipulation or disarticulation, although where there was these could have been soft tissue injuries incurred from the tsunami— perhaps I am drawing a bit of a long bow here and over-speculating! There is a scattering of broken pots and bones (we have seen that before) and also complete pots containing human remains that might have been brought from an entirely different location to the cemetery, which may therefore have acted as a potential catchment area for skeletal remains related to the event. Another nod in favor of these deaths being related to a catastrophic event is that the area was abandoned at this time (around 2800 years ago)—ah ha, so my thoughts about there not being enough room for both the settlement and mass burial at the same time seems to work. Interestingly, the site was reoccupied much later on, and we find that the waste tip (midden) for this new settlement sits directly on top of the burial site. This is a fascinating development because in Polynesian culture such action represents an unusual disrespect for the dead and as such probably indicates a loss of cultural memory about the burial site.

Hmmm—have I made a convincing argument yet? I wonder. Probably not.

Ah yes, context. Over the period that some of the archaeological work was going on at this site located on the south coast of the island of Efate, I had been doing some geological research on other coasts there. The case for the burial site being possibly related to past tsunami inundation is made a little stronger by my findings, which dovetail rather nicely date-wise at least. At two sites on Efate I found geological evidence for the 2800–3000 year old paleotsunami, both on the northwest side of the island, which would have been far more sheltered and more protected from a tsunami emanating from the Tonga Trench. As such, if the tsunami inundated these sites, then the effects on the southern side of the island would have been far more severe. But hey—no one has looked for geological evidence there. Hmmmm.

The arguments in my favor may be less convincing that those for the first case, but here we see again the classic problem we geologists face. Archaeological research has been going on in these islands for well over a century, tsunami research in the region is at best 20 years old. Geological work has not been carried out at these sites and so to a certain extent can be, and usually is, viewed as speculation. Amusingly, it is extremely difficult to put this thinking to bed one way or another because such places are now significant archaeological sites and geologists would be hard pushed to get permission to work there anyway!

Having said that, I do love working in Vanuatu. I have visited several islands in the archipelago, although by no means all 83 or so of them. Apart from Efate where a seemingly endless amount of tsunami work needs to be done but doubtless never will be, there are wonderful places such as Tongoa, which we have talked about in an earlier chapter. Here a story of incest is intimately (excuse the pun) linked with the eruption of the Kuwae volcano.

One of the most ubiquitous things you come across when you visit any of these islands and work with the locals, is Kava. I came across it on Tongoa, Efate, Epi, and Tanna to name but a few of the islands—its use is pervasive. The very word Kava almost sends shivers down my spine.

Some people out there will be wondering why I make such a fuss. If you have never been to the Pacific Islands then you will not have had "real" Kava and have, if at all, experienced it in its powdered form as a sort of relaxant. The real stuff is orders of magnitude more powerful than that, and some more so than others.

I first came across Kava in Fiji back in the 1990s. It was a slightly disturbing muddy color with a pungent taste that is really hard to describe, although I have heard tourists describing it as looking like dirty washing up water and tasting like a wet, slightly gritty mud. This grog is made by pounding sun-dried kava root into a fine powder, straining it, and then mixing it with cold water. It is invariably served in a coconut half-shell and usually drunk on and off for a few hours while socializing until it brings on a numbing and relaxing effect. There is some kind of Kava ceremony in most Polynesian and Melanesian islands, and it doesn't vary too much. The hosts sit cross legged in a line facing the guests who are also lined up sitting on the other side of the Kava bowl and at a respectful distance. One man stirs the Kava and another dips in the half coconut, brings it over to the first person in line, you drink it down in one (that is the politically correct thing to do, and to be honest, the best thing to do), hand the cup back, clap your hands, and the process is repeated with the next person and so on.

The most noticeable effect is a slightly numb tongue or lips, and possibly enough relaxed muscles to make getting up slightly awkward with a little tottering involved from time to time. A nice social event on the tourist calendar.

Kava is in the pepper plant family and is very much a Pacific Island staple, first used by seafarers in either New Guinea or Vanuatu (please note for future reference—these are Melanesian islands, not Polynesian), its use spread out throughout the Pacific as Lapita and later Polynesian peoples moved it around the islands. Depending upon its strength, it has found use as a sedative, anesthetic, and for its euphoria-giving properties. This is not Ecstasy, but it can engender a pleasant and relaxed euphoria-type feeling (for me anyway).

In Melanesia, Kava is traditionally prepared by either chewing, grinding, or pounding the roots of the kava plant. It is then combined with water. This is not a powdered form that has been sun-dried for some time, it is much fresher and much more potent, and often known as the "two days" Kava. A purist, who is invariably not a tourist, will prefer the chewed version since it induces the best effect with the fresh, undried moisture-oozing Kava producing a far stronger drink.

In Vanuatu, only the weaker strains can be exported and must be at least five years old. So if you want to really get to grips with the hard stuff you have to visit Vanuatu or alternatively, if you don't realize that there is a difference in Kava strength and you are carrying out research in Vanuatu you just blunder into it. I have had the pleasure of sampling the "two days" version, but I can one-up myself there and say that I have had one that was more like a "two hours" Kava. This variation of rocket fuel is strictly reserved for special ceremonies.

On Efate I had a pretty good Kava, it was not the special one, but it was so far out of my league (based on the milder powdered Fijian version) that I was both revolted and relaxed at the same time! Not only that but we went to a Kava Bar to drink it. This was by no means the traditional Melanesian way of drinking it, which usually involved sitting cross-legged on the floor (dirt or otherwise) at a Nakamal (traditional meeting place) or some other form of meeting place while the Kava preparation ceremony unfolded in front of you. No, this was a plastic bucket of preprepared Kava sitting in the corner of a decidedly dark, dingy, dirty room behind a bar of sorts. Upon ordering a cup or half-coconut full of Kava a cut-off plastic bottle was plunged into the bucket and your drink was poured. The bucket was refilled by straining the crushed Kava-water mix through something akin to a dirty old t-shirt that in reality probably was exactly that.

Joy. Oh, and I forgot. Kava can be a powerful diuretic, a POWERFUL diuretic.

Hmmm—so many things come to mind, not the least of which was health and safety, but after a couple of them I was usually too relaxed to give a toss. But the taste—argh, it was bitter in comparison to the mild Fijian stuff, but as far as I could tell what it gained in bitterness it also gained even more in effect. Numb lips, numb tongue, and wobbly legs were the order of the day, but it was survivable, especially when the flavor was washed down later by a couple of bottles of the local Tusker beer.

On Tanna though I entered the twilight zone of Kava. This was some years after my Efate experience and as with all such experiences, one tends to filter out the unpleasant parts of the process. I was once again on the search for tsunamis, no mass burials for me then, they hadn't been invented, but I knew that Kava might be on the menu, and if I needed to bond with the locals then it had to be done. I was really only hoping to find a few locally generated tsunamis caused by volcanoes popping off or something like that. Although since Tanna is one of the southernmost islands of Vanuatu, it is not a million miles away from New Caledonia, and between the two of them is the New Hebrides trench—yup, another subduction zone where the Australian and Pacific plates get cozy with each other and tsunamis are spawned. It would be great to find a tsunami or two caused by that subduction zone as well.

Tanna is about 40 km (25 miles) long and 19 km (12 miles) wide with a population of around 29,000 people. It is a wonderful island, the people are incredibly friendly, and it is simply gorgeous. I could tell tales of sleeping in a tree house looking out of the window as the active volcano Mount Yasur popped a bit of lava out as regularly as a metronome about once every 30 seconds, but I won't. I could tell you about our visit to the chief of the John Frum cult where the Kava was even better and the bamboo army equipment had to be seen to be believed, but I won't. What I will tell you about though is the forgotten tsunami and the Kava to end all Kavas.

The forgotten tsunami came like a bolt out of the blue. It was only our second day on the island, and we were wandering around its largest town, Lénakel. When I say wandering that is perhaps too strong a word for it, we were leaning up against our truck, and as the market unfolded before us, we watched people wander by, catching the eye of those we wanted to have a word with. Our guide was a chief and he was keen to introduce us to one of the other chiefs on the island because they had a tale to tell. True to form the chief eventually ambled into view and after a brief conversation in Bislama we walked a short distance up the side street—there was only one, so it was "the" side street. Since the town is clustered around a small bay the side street quite naturally went steeply uphill and inland. After a couple of hundred meters we stopped by a massive Banyan tree that obviously served many purposes—bus stop, meeting place, umbrella, sunshade, lunch spot, the list was almost endless, and it was living up to almost all of those roles when we stopped for a breather. The chief sat down on one of the tree's massive roots and we all selected a comfortable spot to sit and listen to what he had to say.

He said nothing and just smiled.

I think we all had a look of bemusement on our faces, we thought we had come here to listen to pearls of wisdom from this wise old chief, but all he did was smile. It slowly started to dawn on me that he and our guide were in on this joke together but what was the joke? As if he heard me the chief silently turned his head toward the main trunk of the Banyan tree and pointed.

About 2 m (6 ft) up there was a small white sign with a cartoon depiction of waves inundating the land, palm trees leaning inland, and a hill rising up to the right. *"MEMORI MAK BLONG DISASTA"* was the heading and underneath is said "1956: *Wan Saeklon I pusum solwata I kasem long ples ia"*—a rough translation being "a cyclone pushed saltwater all the way up to here" (Figure 8.2).

Holy crap, we were maybe 250 m (800+ ft) inland, but heck, we must have been a good 15 m (50 ft) above sea level or more. That would have been some mother of a cyclone.

There was little more to say, and the chief ambled off, happy to have rattled our cages just a little. Within the hour I was on the super-slow internet trawling through endless historical records of cyclones that had affected Vanuatu. I realize that this was probably not a really cool thing to be doing on such a gorgeous island, but something was niggling me. I didn't remember this event at all. It only took about an hour, but I was correct—no, there had not been a cyclone there in 1956. Let's be fair though, the place had had its fair share of cyclones, so it wouldn't have been unexpected to find

Figure 8.2 Tanna, Vanuatu—the sign on the Banyan tree. *Saeklon*—cyclone or tsunami? (Image: J. Goff)

evidence for one then. Not only that though, but one of the main reasons for our research trip was because months earlier Cyclone Pam had devastated the island, ripping the infrastructure apart and leaving few modern buildings standing. Fascinatingly though, most of the traditionally constructed huts survived. Ni-Vanuatu people are not stupid, they have had thousands of years of being pounded by cyclones, they were OK. Part of the trip was to learn about some of these traditional techniques.

So the nasty wave immortalized at the bus stop Banyan tree was not a cyclone, but it was a *Saeklon*. Back in 1956 there was actually no word to differentiate between a cyclone and a tsunami, it was simply a *Saeklon*. But if it was not a cyclone then where did the tsunami come from? There seemed to be no other memories or records of this event elsewhere so it must have been a local source. The answer is unclear at the moment, but there is a candidate and quite frankly I can't think of any other option. The New Hebrides trench lies a little over a 100 km (60 miles) to the west, but if there had been a big earthquake on that then we would all have known about it. However, about 50 km (30 miles) away in between the trench and Tanna, the seafloor bathymetry showed us what looked like an underwater landslide.

Just like the small underwater landslide in Tonga that tossed up a herd of massive boulders on to the land, these things can generate incredibly large local tsunamis that are completely devastating along a very short length of coast—the 1998 tsunami in Papua New Guinea is another example. Often though, they go unreported, especially in such remote islands. In the absence of any other ideas, this underwater landslide seems like a pretty reasonable source. How often these little underwater landslides happen is completely unknown, all we can say right now is that we probably have a record of one hitting Tanna—it happened in 1956, and modern day Lénakel sits right on top of any geological evidence there might have been for it. At least a forgotten tsunami, or *Saeklon*, has now been remembered in print. It seemed that only the chief was aware of the sign even though it watched over the people of Tanna every day as they went about their business. Memorials of events such as this need to be commemorated annually at least in order to keep them fresh in people's minds. Having said that, the sign on the tree is a start and the fact that it is still there is testament to the chief's endeavors.

We had only walked a short distance uphill from the main street for this revelation, but afterward we drove further uphill, up the same road. It was a mere 30 minutes' drive almost straight up the hill, to get to a village that was well known to most in our party. I was the only one living in ignorance. This was not a tsunami visit, this was a social call. There were longstanding family connections here. The village was about as close as you can get to a Stone Age culture in the Pacific. Think of all modern luxuries and they were not there. There were shorts, the odd t-shirt, and knives, but not many, and a few metal implements but that was it. Their entire village had survived the recent cyclone without the loss of a single hut. The hut design was simple but effective, honed over generations to withstand everything that Mother Nature could throw at it. This design lesson was now being embraced by local architects who were creating modern buildings mirroring the fundamental structure of the huts. This was good news, new light through old windows ensuring the continuity of indigenous knowledge.

While the huts survived, their food crops had been ripped from the ground, and to make matters worse an El Niño drought was setting in. The chief informed us that they were living on their "drought" crops. These were foods that they knew would survive a cyclone, survive a drought, and keep them going until the next growing season. What was the food like we asked? The reply was quite simple an honest "it tastes like crap, but it keeps us alive."

It turned out that as a friend of at least three people from the village I was an honored guest. This was news to me and yet again I had the feeling that discussions were going on behind my back. Oh well, no backing out. The chief and I, with one of my buddies as translator, wandered down to their Nakamal, which consisted of an extensive piece of bare earth a bit like a football pitch with a massive Banyan tree lording over all of it. The chief had a quick word with his boys and they ran off into the bush, machetes in hand. "Go and get the good Kava we have a special guest," he had said apparently. And so a few minutes later they returned, a few kava plants, roots and all, clasped in their hands.

This was the rocket fuel of the Kava world.

We sat down and started to chat. It was one of those bizarre conversations where you realize the huge cultural chasm that separates you, but you could find common ground on the simplest of things—how were the crops? Where do you get your water from? And so on. While he would ask, where do you live? How did you travel here?

Time passed while in the background the boys went about their work. Four separate Kava leaves had been placed on the ground and the original two boys, helped by another two, started picking off bits of Kava root, popping them in their mouths and chewing.

This was the real way to prepare Kava.

They chewed and chewed and then sporadically spat out a wad of masticated Kava root onto one of the leaves. The piles grew like anemic dog turds until there must have been a good handful on each leaf. This wet, gooey substance was then added to water (not much) and while I didn't see the straining process, I assume it took place.

And then there it was, ready to go. The chief and I stood up and half-coconut cups were presented to each of us in turn—him first, he was the chief after all. It was at this moment, the handing over of the coconut that the boy giving me the cup smiled. I gulped—he was missing a few teeth and that mouth had never seen a toothbrush, and he had just chewed his way through a dog turd's worth of Kava root just for me. A small dribble of snot trickled down onto his upper lip as he beckoned for me to follow the chief.

There was no turning back, there was nothing I could do. I could not avoid any of the possible diseases that might come out of this adventure. Putting those thoughts to one side, there were protocols to follow, I listened carefully and followed. The chief drinks first, he takes it all down in one, no sipping, but he keeps back a small amount in his mouth and then sprays it out onto the ground to give back to the earth as thanks. OK, and then it was my turn. Right, down in one but keep a bit. There were so many thoughts going through my mind that I cannot to this day recall what it tasted like, but I did manage to hold a bit back and share it with the earth. Phew—no social faux pas there. The chief then repeated the ceremony with my friend and then we sat, or at least the chief and I did.

We just sat and relaxed. I glanced over at the chief and while I was watching he pulled out the largest knife I have ever seen. This was that Crocodile Dundee scene being played out in the depths of Vanuatu—"call that a knife?" I had absolutely no idea what was going to happen next. Not some weird initiation ceremony? After all, Tanna used to be wall-to-wall cannibalism, maybe up here in the old tribes these traditions continued? But no, he calmly started levering dirt out from under his fingernails. I relaxed. In fact the Kava was kicking in fast so I relaxed a lot. This was mightily strong stuff, forget numb lips and mouth. I knew that in an hour's time we had to leave and head back down the hill to Lénakel and then back to our hotel. Time slowed down, every action and thought seemed to happen in slow motion. I noticed that my friend in all his wisdom, bastard, was leaning up against a tree behind me. How do I stand up? I mustn't make a fool of myself. For the next 40 minutes my mind focused solely on the one task, choreographing my ascent to bipedalism. I have no idea whether people were talking, sleeping, playing cricket, solving the problems of the universe, or just watching me. I had to think through the entire process of how I was going to stand up, the precise moves, how I would balance my weight as I levered myself up and got to my feet. With only five minutes to go until our departure I finally summoned up the courage to make my move. I have no idea how long it took but it worked, no tottering, no talking, just a complete focus on moving back to the car. We were pretty quiet on the way back, not because we were bereft of things to say but because our guide was driving, and he had been surreptitiously sampling the Kava for the past hour. The journey seemed to last a lifetime, but he was fine, hardened by years of "two-minute" Kava. I slept for 12 hours.

That was a terribly long aside, but one of many of the eccentric experiences involved in tsunami hunting in the Pacific. There is a book in them alone.

Moving on as one must I will just take us on a very quick visit to Scotland. There are many great things to do there. Hiking, golf, wearing a skirt, sampling endless malt whiskies, getting eaten alive by midges, the list of activities is almost endless, but who has ever heard of tsunamis up there? This is a very reasonable question that applies as much to Scotland as it does to all of Great Britain. The place doesn't exactly sit anywhere near a boundary between two tectonic plates.

Earthquakes are few and far between, there certainly aren't any active volcanoes up there, but ah ha, they have a near neighbor with an unstable continental slope. The continental slope is the bit between the continental shelf and the abyssal plain. The continental shelf is the part of a continent that is just offshore and goes to a depth of around 100 m (330 ft) underwater. When the sea level was lower during ice ages the outer edge of this shelf was the coastline, all submerged now beneath the sea until the next ice age. The abyssal plain is pretty much as it sounds. Deep and flat, it is an underwater plain that forms the ocean floor between about 3 and 6 km (1.8–3.7 miles) deep. There is quite a lot of it—more than 50% of the Earth's surface. The continental slope therefore tends to start in water depths of around 100 m and drops down to 3 km or more. It can be pretty steep, and that is where things can start to get a little problematic—with steepness comes instability.

Norway is that near neighbor of Scotland, and it was down their continental slope that perhaps the most well-known tsunami generating underwater landslide in the Northern Hemisphere occurred. This happened around 8150 years ago, and it is known as the Storegga slide. The tsunami it spawned was a region-wide catastrophe. The size of the slide was about 20% larger than the whole landmass of Scotland, and the tsunami it caused has so far been traced from Norway to Greenland, western Scotland, north-east England, and Denmark, although doubtless Ireland will get in on it soon and the Faroe Islands. I can almost hear the sigh of relief from every-body: "That's OK, it was ages ago." Ah, well, yes it was ages ago, but it killed people, probably a lot of them because this was the Mesolithic and there were plenty of people around foraging in the islands and coastal areas of the re-gion. But I am more interested in a younger event, one that occurred around 5500 years ago. From the deposits that have been found, this 5500 year old event seems to have been pretty similar to the earlier one and might have come from the same Norwegian source, although it was probably only about half the size with run-ups over 10 m (33 ft) as opposed to over 20 m (66 ft). It is known as the Garth tsunami, because this is the name of one of the places where deposits have been found on the Shetland Islands. There is actually a little debate over where it came from because there is an alternative subma-rine landslide source, the Afen slide, but the jury is still out as to whether this was big enough to produce the tsunami evidence that has been found.

By the time of the Garth tsunami we had moved on from the Mesolithic to the Neolithic. There are many small burial cairns on the Shetland Islands but only one site that has enough intrigue to be of interest.

The Sumburgh cist is at the very southern end of the island called Mainland. This is a cist burial or if you like, a stone lined pit, and it was a mass burial. It contained the mixed remains of something like 27 individuals. As one might expect (he says hopefully) for a mass burial related to a catastrophic event the burial was simple and the individuals in the burial probably consisted of a representative cross section of the population at the time—both sexes, adults, juveniles, and babies. I have read through as much as I can find about this site, BUT just like the Solomon Islands and Vanuatu, I have not visited it. Apparently the skeletal remains show that the people suffered a fair degree of trauma and disease but there was also evidence of fracturing and pitting of the bones that seemed most likely to be associated with gnawing by dogs. Ah ha—they had probably been lying on the ground for some time to allow this to happen.

The cist had been dug into a layer of sand that was overlying the soil at the time the pit was dug. It was noted that that the bodies must have been buried shortly after the deposition of the sands into which the pit had been cut—hmmmmm. A short distance away from this cist is what is known as the West Voe midden. It is overlain by a sand layer that dates to the same time as the one at Sumburgh. OK, hang in there, it might get a little complicated, but I will try my best.

Right then, the West Voe midden is overlain by that same sand, so it is older than the cist burial. This midden or waste dump is made up a lot of marine foods but the important ones for my argument here are the shells—oyster shells, limpets, and mussels (*Mytilus edulis*). Now, once the sand had been deposited on top of it a new midden was started. The sand must have been deposited very rapidly because the radiocarbon dates for the re-established rubbish dump on top of the sand are only very slightly younger. The big difference here is that the new midden is comprised entirely of cockles.

The big question to ask here then is what the hell happened to the environment between the two middens, the one being only very slightly younger than the other and separated by a sand layer? Similar midden changes in other parts of the world have been separated by paleotsunami sediments. There are also dramatic changes to the ecology of the nearshore coastal waters caused by the high-energy tsunami waves rapidly redistributing sediments offshore. In the case of Sumburgh/West Voe this seems likely to be showing an almost instantaneous change from bedrock (oyster shells, limpets, and muscles need rock to grow on) to sand (cockles like sand to grow in).

What of the sand? Well, what we do know is that it is a fining-upward medium sand with microscopic marine beasties in it—this was not put there by the wind.

There we have it. Well not quite. As one of the archaeologists who worked on the Sumburgh site stated, "if these [the sands] were associated with a tsunami, then the implications for human settlement would be significant." It would indeed. It will be interesting to see how this develops in the future. My hats off to the archaeologists who work in these wonderfully remote islands, they are doing a fantastic job, it is a bit like the Chatham Islands without the boulders!

With that we almost have all the building blocks, the strands, in place but first we have to venture to pastures new, to a land where tsunamis are common, where Pisco was born, and where the climate is almost infinitely variable.

9

Strand 8

A Country with Latitude

participants reported that their ancestors possessed understandings that big earthquakes occur every 100 years and that after earthquakes the sea always runs up. This knowledge was directly transmitted through tales or legends by elders. Interestingly, the idea of a cyclic nature of the earthquakes correlates with recent geologic research: four big earthquakes in that zone preceded the 1960 one, with a regularity of approximately 100 years, on 1575, 1657, 1737 and 1837.

E. Kronmüller et al. Exploring indigenous perspectives of an environmental disaster

Welcome to Chile.

Chile is a really unusual country. It looks as if it might have been a mapmaker's mistake. Did they draw their lines a bit thick when they first mapped South America only to find later that when they used a finer nib on their pen an entire country had to be put in to allow for gap between the thick and thin lines? It sits on the western side of South America and extends approximately 4300 km (2700 miles) from its northern boundary with Peru, at latitude 17°30′ S, to the tip of South America at Cape Horn, latitude 56° S. Once you get that far south you are only about 650 km (400 miles) north of Antarctica. The country may be long, but it is narrow. At its widest it reaches a quite reasonable 300+ km (200+ miles) in width but at its narrowest it is a mere 15 km (9 miles)!

Depending upon who you are listening to, Chile can be even longer, reaching up to over 7900 km (about 4900 miles) from the most northern city of Arica to the South Pole. This includes the Chilean Antarctica Territory that they lay claim to and just to add to this confusion, Chile has two geographical centers with monuments to prove it. One is, not surprisingly, about halfway

In Search of Ancient Tsunamis. James Goff, Oxford University Press. © Oxford University Press 2023.
DOI: 10.1093/oso/9780197675984.003.0009

down the coast of Chile as most of us know it, the other which I stumbled across by mistake, is 49 km (30 miles) south of Punta Arenas, the largest and coldest city south of the 46th parallel. Here at a fork in what was then a gravel road heading nowhere is a large white obelisk and plaque . . . and that is it. It might seem just a tad pretentious to claim a vast chunk of Antarctica, but they have a fairly permanent population of a 100 or so people down there who live mostly in the town of Villa Las Estrellas (Town of the Stars). The town boasts an airport, a bank, a school and childcare center, a hospital, a supermarket, mobile telephone facilities, and TV reception. Hmmm, better than my village.

We have already come across Easter Island/*Rapa Nui*, well that's Chilean too, as are the Juan Fernandez Islands made famous by Daniel Dafoe's novel Robinson Crusoe. While he put the island elsewhere, the story is largely based on the life of Alexander Selkirk, a Scottish castaway who lived for four years on one of the Juan Fernandez Islands that was called Más a Tierra but was renamed Robinson Crusoe Island in 1966. Both these island archipelagos have their own unique tsunami stories, but we will not get distracted by those, our focus is on mainland Chile.

Given the vast latitudinal range of Chile it is hardly surprising that it also has an incredible climatic range. It is home to the world's driest desert— the Atacama—in the north, and then moves through a rather balmy Mediterranean climate in the center down to the gorgeous Alpine and sub- Antarctic climates of the south, with glaciers, fjords, lakes, and some glorious mountain ranges. Of course if you keep going you end up in the polar climate of Antarctica, the world's largest, albeit frozen, desert. So the entire country is bookended by deserts, both of which have had run ins with tsunamis in the past.

Needless to say, given the huge variety of landscape and climate, hunting for tsunamis in Chile requires every single possible strand of information you can get your hands on. If we are talking about wearing new glasses, then this requires not one new pair but a veritable optician's worth of different types. A tsunami deposit in the south of the country is almost unrecogniz- able from one found in the north. Perhaps the most telling point is that from about the middle of the country southward (based upon the first monu- ment), the tsunami deposits are gorgeous textbook examples. To be honest, you would have to be blind not to see them—peaty wetland soils overlain by a lovely thick sand layer that fines-upward and is then overlain by peat again. Inside the sand you can find all the key food groups—marine microfossils, geochemical signatures for saltwater, bits of underlying peat that has been ripped-up and incorporated into the deposit, etc., etc., etc. Beautiful, and in many places this pattern is repeated vertically several times over like a German schichttorte or multi-layer cake (Plate 9.1). And just to add to the ease of the geology they have their legends and superstitions too.

The Lafkenche (people of the coast), a subgroup of Chile's indigenous Mapuche people, live on the coast near the epicenter of the 1960 earthquake (and tsunami). This massive Magnitude 9.5 earthquake generated a devastating tsunami, but they were not at all surprised. Their *Machi* (shaman) had dreamt about this event some time before it happened and had predicted it, but this dream was just part of a series of complex indigenous knowledges and beliefs. For example, they had a tradition that these major events happen in a cycle about once every 100 years. While this had been held in the indigenous knowledge for generations, modern science only recently caught up by finding that the 1960 was preceded by similar events in the region in AD 1837, AD 1737, AD 1657, and AD 1575. This rather nicely makes the point that it always helps to ask the locals first to find out what they know.

Having said that, not all Mapuche beliefs are quite as benevolent. The catastrophic 1960 earthquake and tsunami was the catalyst for the ritual human sacrifice of a 5-year-old orphan boy in the isolated coastal village of Collileufu, again not far from the earthquake epicenter. This was also an indigenous Mapuche community but on this occasion both the traditional healer and religious leader insisted that there needed to be a human sacrifice in order to calm the land and the sea. The inevitable upshot of this heinous crime was that these two (men) were jailed for murder. The judge ruled that all of those involved "acted without free will, driven by an irresistible natural force of ancestral tradition." Hmmm, OK, so indigenous knowledge is a powerful tool in helping modern scientists understand the environment, but indigenous superstition is another beast altogether.

This part of southern Chile provides a rich vein of geological and human data that has revolutionized our understanding of the tsunami hazard in the region. It is where many tsunami scientists have cut their teeth, studying and analyzing these layers and traditions, and yes, in many cases taking away this expertise and using it elsewhere. This is to be commended as part of their training and knowledge, after all tsunamis don't just happen in places where Mother Nature is kind to researchers. Many researchers have fallen into what I would call the Wegmann Trap, staring down their own little tube and focusing more and more on what they learned and where they learned it, picking it apart and searching out even more details from their own personal layer cake experiences. That can be a big problem because places like northern Chile are not so kind.

And this is where we briefly recap what was said at the end of Chapter 1.

Northern Chile is dominated by the Atacama Desert, a strip of land, more than 1600 km (1000 miles) long, squeezed between the coast and the Andes. The Atacama Desert is the driest non-polar desert in the world, with an area of around 105,000 km² (40,500 sq miles) composed almost entirely of stones and sand. Chile was invaded, colonized, conquered, and crushed by the Spanish in the 16th century. Or at least the colonizers got a little bit beyond Chiloé, which is just south of the epicenter of the 1960 earthquake. While it is likely that the Spanish met some stiff opposition from the indigenous Mapuche who had had time to organize themselves by then, it is generally thought that the Spanish lacked any incentives to go further south—it was bloody cold!

In the north though, things were different.

Arica, Chile's most northerly city, is located on the coastal edge of the Atacama Desert. It is there because it is at the mouth of the Azapa River, a necessary requirement for a stable water supply and a lush(ish) vegetation. It was always destined to be an oasis among the dry hell around it. Founded by the Spanish as Villa de San Marcos de Arica in AD 1541 on the site of a much earlier settlement, it was Peruvian until AD 1879 when it was captured by the Chileans (with the help of the British) during the Saltpeter War. It has remained Chilean ever since. Prior to any of this historically documented activity it had also been the home of the Chinchorro hunter-gatherer culture from at least 9000 years ago, and probably as much as 12,000 years ago. As a culture they are most famous for their intricate mummification process. The importance of Chinchorro mummies is gradually being recognized, after all they are much older than the mummies of ancient Egypt. Sadly though, I have seen numerous Chinchorro mummies simply sliding down the side of road cuts. They may be culturally important but modern development seems to be even more so. The Chinchorro culture pretty much ended around 3500–4000 years ago, overwhelmed by more recent arrivals—the Tiwanaku Empire from the Andean Plateau who were farmers and agriculture became the name of the game.

I probably didn't need to give such a long history lesson, but it is important for two reasons.

First of all, Arica and the Atacama Desert coastline to the south has been occupied by humans for at least 12,000 years. This may seem odd given that, except for the odd river, it is probably the driest place on earth. However, human life has not only been sustained here, but it has done rather well thanks to the offshore Humboldt Current and the cold water it brings from the south. This cold water wells up to the sea surface bringing with it a massive amount of nutrients, which in turn supports a huge biomass of phytoplankton, which in turn brings in the fish and bigger players. Fishing was an important activity from the very early days of human settlement here. Not only that, but in some parts of the Atacama Desert coastline a marine fog known locally as the *Camanchaca* marches inland for a short distance providing a few parts of the coastal strip with sufficient moisture for plants to grow. Add to this the mixed blessings of often catastrophic El Niño rains, and you start to get a sense that while life here is precarious, it can work. El Niño events occur about once every five to seven years or so when warmer waters from the north push the Humboldt Current out of the way and suppress it. This is bad for fishing since stocks plummet, but the warmer air flow across the region can bring dramatic changes in rainfall. On occasion it can change from an average of less than 4 mm per year to almost a decade's worth of rain falling overnight. The bad side of this is that it can cause massive flooding. Normally dry canyons or *quebradas* that come off the hills and down to the coast almost instantly fill with devastating slurries of water and sediment inundating the sparsely populated coastal strip often leading to at least few deaths even today. Is this yet another reason not to live there? Indeed not, because El Niño rains can also lead to the sudden germination of millions of seeds and a phenomenon known as *desierto florido* (desert in bloom). This may only last for two or three months, but it supercharges the environment and food is plentiful. And then the area returns to the normal struggle, the Humboldt Current eventually re-establishes itself, and fishing is flavor of the day again.

Second, as if the climatic and ecological extremes are not enough, living along this coastline means that humans have been exposed to numerous massive earthquakes and tsunamis for the entire time that they have lived in the region. Immediately offshore, two tectonic plates meet—the subduction contact between the Nazca and South American plates. In the time since the records of such events have started to be written down we know of earthquakes and tsunamis in AD 1615, AD 1768, AD 1786, AD 1868, AD 1877, 1967, 2007, and 2014, and that is just in the northern part of the Atacama

Desert, north of the Mejillones Peninsula. To the south, the picture is different, not in the sense of being any quieter but the dates are all off, with events in AD 1796, AD 1819, 1918, 1922, 1946, 1966, and 1995. Some were bigger than others, but the lengths of the fault that actually moved never seem to have crossed a major sticking point that lurks just off the Mejillones Peninsula.

For example, the Magnitude 8.8 or so, AD 1877 Iquique earthquake caused about 500 km (310 miles) of the fault to move between Pisagua-Iquique and Mejillones. The tsunami was a bad one. Locally it was 15–20 m (50–65 ft) high but it went Pacific-wide as well with waves up to 3 m (10 ft) high inundating the distant shores of Japan. It is not surprising then that numerous tsunami researchers from around the world have wandered down this chunk of the Atacama Desert coast hoping to find their paleotsunami El Dorado, applying well-developed geological techniques . . . and until very recently they came up blank almost every time. In Peru, immediately to the north of Chile, researchers studying how well recent tsunami deposits were preserved in an almost identical coastal desert landscape found that things were a bit grim. Three moderately large tsunamis had inundated the Peruvian coast between 1996 and 2007, and their preservation along this arid coastline varied a lot. In one instance all traces of the tsunami had been removed or reworked by flash floods (El Niño bless it, cleaning up the mess) and stormy ocean waves. In another, the wind had removed all the fine sediment, and in another storm waves again eroded the evidence.

Grim news indeed—a tsunami deposit on a desert coast seemed to have a lifespan of only 10–15 years. Despite historical accounts of the massive tsunami in AD 1877, no geological evidence of either this event or any of its predecessors has been found recorded along the Atacama Desert coastline (well, that's not entirely true, but our paper hasn't been submitted yet . . .). How on earth could anyone be expected to find anything older? And so to a large extent, El Dorado faded from the wish list. But let's cut to the chase and revisit the email that landed on my desk and made this entire venture possible: "We are currently working on the hypothesis of a mega earthquake and possible tsunami occurring in northern Chile around 4000 years ago." Could I help?

How on earth could you even think about the possibility of trying to even look for such a deposit when these things seem to have a lifespan of about a decade or so?

It is important to remember though that this book is not about the search for a 4000-year-old paleotsunami. It is, rather, a look at what we do to try and find out whether something like this has happened in the past. What lines of evidence can be used and what new thinking can be brought to bear on the problems of working in such extreme conditions?

In a sense this is what the toolbox and the nexus of individual strands is all about. It is all very well pottering around your local backyard sounding intelligent, but if you really want to test yourself then go to a new place where you have never worked before, with the worst possible preservation potential in the world—and go for it. After all, we know that lots of tsunamis have struck that coastline in the recent past and we know damn well that they will happen again in the future. So a scientist who is supposed to know about these things cannot simply reply to the email, saying, "sorry, we won't find anything." I do what I do because I want to save people's lives, and to do that we must learn as much as we can about past events—there could be no thought of turning such an offer down. There is, however, the possibility of failure and of history repeating itself. After all up until that point no one else had found anything of substance.

One may say something like, "who gives a damn, hardly anyone lives there anyway?" One might say that, but people do live there, and actually quite a few. The port city of Iquique is on the edge of the Atacama Desert, its port facilities and high-rise condos hugging the narrow, low-lying coastline. The city developed during the heyday of saltpetre mining in the Atacama Desert in the 19th century, and was then in Peru. But Chile captured in it the War of the Pacific and now copper mining is all the rage. Oh yes—it has a population of well over 190,000 people, most of whom have nowhere to run when a tsunami strikes. So I guess you could say that there is a very real need to know the nature of the tsunami beast, if that is possible.

A hostile environment where the geology and climate are just completely against you. What could go wrong?

Well, judging by the tsunami work that had been done before along the Atacama Desert coast it sounded like half of the wonderfully developed toolbox of techniques—the strands—could simply be thrown out of the window because they couldn't possibly work in a desert. Couple that with the fact I would be joining an entirely new team of colleagues many of whom were archaeologists who also spoke a different language, it would be fair to say that I would also be most definitely stepping out of my comfort zone. So yeah, plenty could go wrong.

It may have been the case that this could all go belly up and fail, but at least I made sure to line up all the weasel words possible to give me a "get out of jail free" excuse. However, to be honest, this was possibly the most exciting professional adventure of my career. A brilliant test for everything I had accumulated like barnacles on my hull over the years. This would not be a simple go in, dig a few holes, pontificate a bit, take a lot of samples, have a great time, and return to base camp study trip. This was the ultimate. No staring down any tubes here, blinkered to what others were thinking and doing. This would require every little dredged up brain cell of memory to be shared with the ideas of others, who had spent a lifetime working in this environment.

To suggest that I had not been to this neck of the woods before though would be incorrect. Almost a decade earlier I had received funding to go to Northern Chile to check out apparent paleotsunami deposits left there by the Eltanin tsunami that was caused by a massive asteroid falling into the very southeastern part of the South Pacific Ocean. I have actually written about this one before in an earlier book chapter, but it warrants a quick recap. In reality it warrants an entire book going all the way from the history of the ship that was used to find signs of the asteroid through to the possibly (probably?) massive global environment impact it had—perhaps that's the next book.

To recap, the Eltanin asteroid hit the deep ocean 2.51 ± 0.07 million years ago, that's two and a half million give or take a few hundred thousand years. Now that may seem a long time ago, but it is not only the most recent big asteroid impact we know about (a big asteroid is somewhere between 1 and 4 km, 0.6–2.5 miles, in diameter), it is also the only known deep ocean impact—it fell into the 5 km (3+ miles) deep Pacific Ocean. Hang on I hear you cry, what about Chicxulub, the one that did for the dinosaurs? Well, that was a longggg time ago, some 66 million years ago or so and while it was huge, about 12 km (7.5 miles) or more in diameter, it fell into a moderately shallow part of the sea. You will notice that there are a lot of vague words here—"some" 66 million years ago or so, "about" 12 km or even more, and so "moderately" shallow therefore equates to somewhere between 100–500 m (330–1600 ft) deep or more. There is quite a lot of variability in all these figures, but we seem to focus a lot on this dinosaur-killing beast and what it did, which I suppose is not surprising. BUT, the Eltanin is so recent that we have taken our eyes of this one—2.51 ± 0.07 million years ago is pretty damn precise. AND as if that was not enough. the last decent major climatic change that marked the beginning of what is known as the Quaternary period started . . . 2.58 million years ago. Hmmmm—now is that a mere coincidence? I will not go into this in great detail. For that you can read one of my other books, *Tsunami—The World's Greatest Waves*. I devote a whole chapter to the Eltanin in it and join up a bunch of ecological dots that seem to point to 36% of all life on Earth going extinct at that time, including the megashark (although if you believe Hollywood it somehow survived in order to allow them to make a disaster movie about it). This may seem like an outrageous suggestion to make, and I am sure that some colleagues have probably assigned me a padded cell somewhere, but the big problem as I see it is that the Eltanin asteroid impact was such a recent event that we can't see the wood/forest for the trees, a comment I have made before about work elsewhere in the world. We have so much scientific data for what went on around that time that it is difficult to sort it all out, so I decided to put my neck on the scientific chopping block just to see what might come of it . . . nothing much yet.

The interest with the Eltanin tsunami is that many researchers have "found" deposits that purportedly relate to this event while others have "not found" such evidence at the same sites. Some may be tsunami deposits, some may not be. It is a mess. Just to add to the confusion, no impact crater has been found, just a bunch of disturbed sediments and an iridium layer. (Iridium is from the platinum group of metals—white, hard, brittle, with a very high melting point—hence when a massive, fast-moving lump of rock smashes into the deep ocean you have the classic case of an—almost—irresistible force meeting an immovable object—the ocean acts like a solid in these cases—and iridium is released as the asteroid is pulverized into dust.) The possible size of the asteroid was calculated by the amount of iridium found in the disturbed sediments. There was probably no crater on the seafloor though because the asteroid fell into an ocean that was about 5 km (3 miles) deep, and at a maximum size of 4 km (2.5 miles) would not have reached the bottom before being obliterated. But the entire process could have generated a massive tsunami.

The size of the possible tsunami has been calculated. In theory, the asteroid could have blasted the water off the ocean floor for at least 60 km (37 miles) around the impact site and created a wave over 200 m (660 ft) high on the southern end of Chile, it would have smashed into Antarctica, 35 m (115 ft) waves would have impacted Tasmania, Fiji, and Central America, and 60 m (200 ft) waves would have washed up on New Zealand's shores.

Right, OK, that was the original thinking, a classic "Deep Impact" Hollywood thriller. More recent work indicates that the wave would have been huge at the impact point, several hundred meters high, a true mega-tsunami, but because an asteroid impact is a point source the energy radiates out from this point and as it expands across a wider and wider area this energy is rapidly dissipated. There is another key wrinkle here as well. This rapid decay in wave size is helped by what is known in the trade as the Van Dorn effect—these types of deep-impact-generated waves go rushing toward the coast, hit the edge of the relatively shallow continental shelf, and break a long way offshore with a lot of their energy also being reflected back out to sea. Essentially, they get a lot of the stuffing knocked out of them long before they can do much damage. However, even allowing for these massive drains on the tsunami's energy, it is still reckoned that a wave 20–30 m (65–100 ft) high would have struck the Chilean coast.

OK, so even if it was *only* 20–30 m (65–100 ft) high it warranted some research to check out what had been found and what else might be around. Was it as bad as has been made out? After all, depending upon who you considered to have got the math right the wave had probably been 20–200 m (65–650 ft) high when it hit the coast. How far did it go uphill and inland after that?

Needless to say, a visit to northern Chile to search solely for Eltanin deposits seemed like a bit of a wasted opportunity. And so began a two-fold operation—start in the very far north of Chile and work down the coast as far as the first apparent Eltanin deposit that had been reported by anybody, check out the Eltanin evidence but also take the opportunity to look for more recent tsunamis as well. After all no one else had managed to find anything, so it was worth a look.

Enter Arica, the first port of call, as far north as you can get. No Eltanin deposits have been reported from here, but it was the most northerly airport AND the site of the devastating AD 1868 and AD 1877 tsunamis, so it was always going to be worth a look. On the first day we managed to visit the site of what remained of the Peruvian warship, the *Americana*. In the AD 1868 tsunami this ship was picked up and carried inland, losing her captain and some 85 crew members. There wasn't much to see, just the rusting iron ribs of the bows reaching out of the ground as if she were drowning in sand and crying out for help (Figure 9.1). The wreck was found when developers moved in to build a new block of flats. Work was halted because of the historical significance of the find. I would have stopped it because, duh, the building would have been in the tsunami inundation zone, but I guess no one thought of that when all the neighboring buildings went up.

One of those chance in million things happened next.

Figure 9.1 Rusting iron ribs of the Peruvian warship *Americana*. In the background is the Morro de Arica, 139 meters above sea level, the last line of defense for Peruvian troops during the War of the Pacific (1879–1883). (Image: J. Goff)

Probably only some ridiculously focused tsunami researcher who can never switch off would have seen it (who, me?). We were driving back along the edge of what would have been the old sea "cliff" at the time of the AD 1868 event with a flat expanse of coastal wetland (well, what counted for a wetland there anyway) seaward of it. This took us pretty much through the middle of the city. On the seaward side of the cliff were the abandoned construction works of the Parque Centenario, a proposed leisure park for tourists and locals alike. However, work was shut down in 2011 for reasons I will not go into, but it involved money. Recently though new life is slowly being breathed back into the project, and it will hopefully become a lovely public greenspace in the middle of the city, maybe not as ambitious as the original plan but not just a failed construction site either. What we saw were half-finished fountains and water features, lots of sand, a few desultory trees, and large amphitheater-like holes with half-completed walls and reinforcing bars sticking up out of the concrete. I glanced down at the carnage and out of the corner of my eye I thought I saw a few tell-tale layers of unusual deposits in the side of a trench behind one of the half-completed walls (Figure 9.2). It was a good 100 m (300+ ft) away so it might have been nothing, but heck a

Figure 9.2 Parque Centenario, Arica, Chile. The layer cake is there in this dry and dusty building site. Not quite the same thing as seen in the south, but it has fine sand layers and a little organic material separating them. (Image: J. Goff)

quick detour, a few quiet words in Chilean to people on the site, and we were there. As quick visits go, I was pretty convinced that we might have found three or four tsunami deposits—the kind of ones that no one else seemed to have been able to find before. What a great start to the trip, although sadly no subsequent funding and no time meant that the Parque Centenario site lay in scientific abeyance until nearly a decade later, but it was great to know that I had the right pair of glasses on.

Based upon a series of rough guesses and calculations, the Parque Centenario must have been around about the spot where the famous USS *Wateree* ended up after the AD 1868 tsunami carried it inland and then the AD 1877 one carried it out toward the sea again. This would be a cool place to do more work. It was therefore a great thrill to revisit the site years later, unvisited by any tsunami researchers in the interim (why not?), to do more work. Samples were taken, work was carried out, and results will be out soon—promise (yes, we have found the AD 1868 and AD 1877 tsunamis). The real thrill of this second trip came when the local archaeologist took us a short way across the park to show us the recently discovered last resting place of the USS *Wateree*! It was not much to look at, a small piece of rusting iron protruding from the side of an archaeological trench, but to be in the physical presence, however meager, of such an iconic ship was truly remarkable.

My word, another long tangent, and it isn't over yet.

Finding tsunami deposits related to historical documented (and possibly older) events in the middle of Arica was quite the coup, but as we worked our way south along the coastline of the Atacama Desert several other sites came to light as well. Not many, because much of the coast is either steep cliffs, collapsing steep cliffs, or narrow coastal strip with lots of debris from . . . collapsing steep cliffs. However, it was always worth a search wherever we found a river mouth, dry or otherwise, at the coast, and also checking out those places where there was historical documentation for a nasty tsunami or two. Having said that, this was exactly what all our predecessors had done, so this could have just been a wild goose chase while driving between Eltanin sites.

I want to say something like "imagine our surprise when . . . " or perhaps a more sporting analogy like "against the run of play we were lucky . . . ," but to be honest this was not a competition or even checking up on work done by our predecessors. We were just looking and doing what we always did—taking the toolbox of identifying criteria coupled with all of those "new glasses" ideas that were always bouncing around inside our heads out for a spin along the coast.

We found several exciting sites, and definitely a few tsunami deposits, but we couldn't date them at the time or really do much more than go "oooh, look at that." However, as a reconnaissance mission it was very successful. As we headed further south though there was a sense of growing discomfort. How come we were finding things at exactly the same places our colleagues had looked before but had found nothing? This unease started a mere 20 km (12 miles) south of Arica at a spot called Caleta Vitor. Here not just one site got us excited, but two. Fair enough, each of them were slightly difficult to access. To the north, a low tide scramble over the rocky coast was required, pinned between high cliffs on one side and a raging sea on the other. But it was evident, even just standing in the small sandy bay that opened up 50 or so meters along the coast, that there was something there. It was not as if there was a lovely tsunami deposit sitting on top of the sand. But at a ridiculously difficult to access height above the beach there was a discontinuous line of marine sand and shells perched precariously among the rocks and squeezed into nooks and crannies (Figure 9.3). It was so different to the surrounding rocks that even though it seemed to be on its last legs and barely clinging on to its precipitous home, it was clearly saying "OK, so how do you think I got here then?"

Figure 9.3 Caleta Vitor—it's up here somewhere! (Image: J. Goff)

Pavlov has a lot to answer for. He may have trained dogs to salivate when the dinner bell rang, but we were trained to sample sediments if they looked interesting, even if we had no time or money to do anything else with them. So we did. Perching 20 m (65 ft) or so above the beach on a perilously unstable slope may be some people's idea of fun, though not mine, I get vertigo. But we got the samples, took the photos, and moved on. The tsunami fest continued. At the southern end of Caleta Vitor we found a coastal cave perched some 10 m (30+ ft) above sea level. The cave floor was filled with a deep deposit of fine wind-blown sediment whose quiet and gradual accumulation in this sheltered little cul-de-sac had been rudely interrupted at least twice by some very high energy waves that brought shells, gravel, sand, and what had probably once been seaweed up into this cavern of no escape. Once in there it could not get out and so settled out in a lovely fining-up layer cake sequence. Were these tsunamis or just very bad storms?

It is always tempting to lean in favor of what you are searching for, but until these cave deposits have been more thoroughly studied the best I can say is that they are probably related to tsunamis. The arguments here may seem self-serving, but I don't think so. First of all, the wind-blown deposits seem to represent a significant length of time—hundreds to thousands of years. Second, we only found two such deposits, which suggests that the cave is so high it can only be reached by the most extreme of events. The material we found in the cave had been deposited out of suspension in a nice, orderly fashion. It would seem reasonable to think that during a storm large enough to get this high, there would have been several waves able to toss stuff into the cave, the last wave disturbing that which had been deposited before. This was not the case. The same argument might be made for a tsunami, but inundation by a single big wave out of a group of smaller ones is more likely in a tsunami. The coast has suffered both large tsunamis and massive El Niño driven storm events, but my gut feeling is that these are indeed tsunami deposits—because they just feel right. I would expect even moderately small storms to toss up the odd few bits and pieces this high, but there is nothing to show that has happened in which case it would seem that a different type of wave is needed to do the job.

Well, I still have the samples, and no, no one else has done anything since we looked . . . sigh.

These types of findings continued to crop up as we moved south. At Caleta Camarones there was a wetland, would you believe, on the coast of the driest desert in the world. Another 150 km (90 miles) further south near Rio Loa there was a deposit sitting on top of a marine terrace—an old beach, if you like—that had been lifted up by an earlier earthquake. About 80 km (50 miles) south of there, we hit Cobija.

Now, Cobija made its first appearance under Strand 4. It has had a rough 250 years or so of history, let alone prehistory. It used to be known as Puerto La Mar, which can be loosely translated as Seaport. The name is not surprising really since it was the first significant Pacific Ocean port for the landlocked country of Bolivia. This was incredibly important because as early as AD 1825 Cobija was the main port for exporting silver from the Potosí silver mine. To say that the Potosí silver mine was big is something of an understatement. It was founded by the Spanish in AD 1545 and was soon producing an estimated 60% of all the silver in the world. With that came huge wealth and a complex supply chain. The mine sits at an altitude a little over 4000 m (over 13,000 ft) which made access to and from it something of a logistical nightmare. For example, the bulk of the silver was taken west by llama or mule train to the Pacific Ocean, then north by ship to Panama City, then mule train again but this time east across the Panama Isthmus, and then by ship to Spain on the Spanish treasure flotillas. For that amount of effort, it had to be worth it.

Cobija was not to reign supreme for long though, the city was wiped out by tsunamis in AD 1868 and AD 1877, and then its ownership changed from Peru (Bolivia had an agreement with them to use the port) to Chile in AD 1884. Unfortunately, Bolivia sided with Peru against Chile in the War of the Pacific, and Chile won. More ignominy awaited, as if enough salt hadn't been rubbed into the wound, because Cobija was replaced in 1907 as a port by Antofagasta some 125 km (75 miles) to the south. Cobija was abandoned, and all that remains of the city today are the barest skeletons of once proud buildings sitting on a low hill above the ramshackle houses lived in by those still carrying out the traditional maritime way of life of fishing and seaweed collection.

What more ignominy could have befallen poor old Cobija in the prewritten past? Now that is an interesting question that sits within the context of what we know about the recent past. Cobija, or at least the bit of coastline around it, has been, and doubtless always will be, an important point along the coast for access to the sea by boat. After all, Cobija was destroyed three times, twice naturally by tsunamis and once by humans who moved everything away in 1907. And yet every time people have returned to Cobija, either rebuilding their port or coming back to continue their traditional maritime existence. It is therefore not surprising to know that we can trace human occupation of this site back thousands of years—at least 7000 years and undoubtedly far more. So while Cobija is now a shadow of its former self, it has been popular for a long time, and it is therefore only to be expected that prehistoric occupiers of this address had their run-ins with tsunamis as well.

As an example, during what is known in Chilean archaeology as the Formative Period which began about 3000 years ago, we see that Cobija or more specifically the Cobija 10 burial site, marks the southernmost limit of a tumuli/burial mound tradition. This was first identified way up north in Arica but it seems to have started a little later down at Cobija, around 2600 years ago. When you are fixated on tsunamis and the high death tolls we have experienced in the modern era, it is difficult not to start thinking mass burials as soon as someone uses terms such as tumuli or burial mounds. Given our somewhat limited knowledge of prehistoric tsunamis along this coast it is seductively exciting to realize that the Formative Period here ended around (there's that word again) 1600 years ago, about (another weasel word) the same time as the coast was struck by a tsunami roughly 1500 years ago or so. Hmmmmm. Having said that, we really only know about four events in prehistory around here—about 950, 1300, 1500, and 3800 years ago. While 3800 years ago rings a bell and will continue to resonate for a while yet, 950 and 1300 years ago seem to float on their own with no obvious linkage to human disturbance. Although having said that, two regional cultures, the Wari who lived in the highlands and the Tiwanaku who had moved down to the coast, seem to come to an end around 950 years ago—why? No idea, but once again the intriguing carrot of the paleotsunami hangs over our heads. Who knows what further potential convergences will be dug up by future geological and archaeological research?

Before we leave Cobija we need to briefly wander back to the Potosí silver mine because it features in another important Peru-Chile debate. At the end of the last chapter I made what many might think of as a sacrilegious statement, "With that we almost have all the building blocks, the strands, in place but first we have to venture to pastures new, to a land where tsunamis are common, where Pisco was born."

Pisco is an amber colored brandy made by distilling fermented grape juice into a high-proof spirit and was one of the good things to come out of the brutal Spanish conquest of Peru and Chile. It was conceived early on in the Spanish settlement when the settlers needed a domestic alternative to Orujo, a type of brandy imported from Spain. The vines arrived from the Canary Islands and vineyards took off. They didn't make brandy to start with though, but wine. The missionaries were the first in on the scene and used the grapes to produce sacramental wines, but since wine was really expensive to bring all the way from Spain and it had usually gone off by the time it arrived, suitable areas for vineyards were rapidly brought into production in Peru. In many ways this was done just as much to keep the locals in good spirits (no pun intended) as it was to feed the voracious thirst of the miners in the Potosí silver mine. A similar scenario unfolded in Chile to the south, with the same thirsty markets.

Wine was produced in huge quantities within the Peruvian colony alone. By the end of the 16th century production was as high as 80 million liters (about 17.5 million gallons) of wine a year. Not surprisingly, Spain didn't like this and so in AD 1595 King Philip II attempted to protect the Spanish wine industry by prohibiting the establishment of any more commercial vineyards in the colonies. And, as in many a rebellious colony, the law was ignored. Apart from anything else this was a key part of the fledgling economy.

In AD 1600 though, Mother Nature had her first shot at shaking things up a bit. The Huaynaputina volcano erupted, devastating Peru's wine production, which was forced to move north to pastures new, and off it went again. Spain continued to be incredibly annoyed at the recalcitrance of its Peruvian (and Chilean) colony and ultimately banned Peru's exports of wine completely in AD 1641.

What was good for Peru was also good for Chile and so faced with this slight wrinkle to their wine production they turned to distillation and Pisco was born. It proved to be a great success in three ways—it provided a wonderful brandy for the local market, it could be transported great distances without going off like wine and so fed a growing international demand, and finally the Potosí miners had a reliable tipple that could withstand the rigors of transport through the Atacama Desert.

Life was good until AD 1687 when Mother Nature did her bit again with a massive earthquake that destroyed many vineyards and their stocks in Peru, and to cap it all, the resulting tsunami washed away the local shipping port.

With no way to ship what little alcohol was left producers in the area distilled what they could. It was at this point that the Chilean producers who had avoided this calamity to the north stepped up to the plate and filled the gap in the Pisco market.

This ongoing tussle over alcohol between Peru and Chile continued over the years, with Peru generally coming off worse. Earthquakes and tsunamis in AD 1868 and AD 1877 struck the vineyards of Peru and their ports again. The War of the Pacific (AD 1879–1883) saw Chile take over a large swath of Peruvian vineyards. But possibly the most significant unexpected blow came in AD 1888 with the arrival of phylloxera in Peru. Phylloxera is a microscopic louse or aphid that lives on and eats the roots of grapes. In the late 19th century the phylloxera epidemic destroyed most of the vineyards in Europe and because the Peruvians also used the European wine grape *Vitis vinifera*, they were also devastated. Ironically, if they had used wine grape species from their neighbors in North America they would have been far less affected, but in the 16th century when the vineyards were started they didn't even know about the North American grapes. To the south of Peru, Chile's relative geographic isolation saved it from the blight and so yet again Chile came out on top.

Ultimately there is not a lot of love lost between the Pisco producers of these two countries. It is an age-old battle between Chile and Peru over a brandy made in the driest desert in the world. In reality at the time the vineyards started up there wasn't really a significant border between the two colonies and it has only become more of a heated debate as borders have shifted over time and national pride has started beating its chest.

Perhaps Peru has had the last laugh though, finally setting up their own Pisco denomination of origin in 1991. This was a shrewd move since in 2013 the European Union recognized Peru as the birthplace of Pisco. Chile can still sell it in Europe, but they do not have any claim to its geographical origin. So Chile won the battles, but Peru won the war, which to be honest is probably fair—after all, the drink called Pisco is quite likely named after the Peruvian port of Pisco where it is thought to originate from.

Having said all of that, the famous cocktail, Pisco Sour, is made differently in both countries. The battle or the war is now being fought over which is best, and I suspect that this will continue for years to come. Which is best? Go there and find out. Don't wander into a fancy bar in New York or London or wherever, it is best to try them in their respective countries—the ambiance, the persuasive explanations of the serving staff, and the wide variety of Piscos available make it an experience you will never forget . . . unless you have too many.

Was Pisco born in Chile or Peru? The answer seems to be that it was probably born at almost exactly the same time in both countries, although I am sure there is an argument somewhere that gives one of the countries a season or two's start over the other. Who cares—it is a wonderful drink to enjoy after a hard day in the field.

Now it is time to get back on track once more and onward from Cobija to Hornitos, a mere 30 km (18 miles) or so to the south. And it is time to finally say something about the Eltanin tsunami, the original raison d'être for these South American rambles.

The findings of researchers working at Hornitos over 20 years ago make salivatory reading for a tsunami scientist. Deposits up to 10 m (30+ ft) thick and between 12 to 20 m (40–65 ft) above sea level. Hornitos marks the northern limit of a suite of deposits that include 5 m (16 ft) diameter boulders weighing 200 metric tons (about the same in imperial measures) and a cluster of up to 20 remarkably intact baleen whales found up to 800 m (half a mile) from the current coastline. When talked about as a group these findings make a compelling case for the Eltanin tsunami, but when examined individually in detail they start to resemble a house of cards that might collapse at any second. Many sites have been documented over 1600 km (1000 miles) of Chilean coastline. Could the Eltanin tsunami have affected such a great length of coastline with such devastating effect? The answer to that question is probably yes and no. Some of the apparent tsunami deposits such as those at Hornitos have been questioned, disproven, and/or reassigned to more prosaic interpretations BUT, and this is the key point, many have not. On the flip side, other deposits that could be tsunami-related have been given entirely different interpretations without any consideration of a possible tsunami origin . . . ugh, what a mess.

While some of the tsunami interpretations may be correct it is difficult to be entirely convinced. Of course a non-tsunami interpretation is likely to be the most acceptable explanation to most researchers since tsunamis large enough to leave such massive deposits are incredibly rare on any coastline and identifying them is not within the sphere of expertise of most scholars— even those who study tsunamis!

In this context it is important to recall that the study of tsunami geology is a relatively young and minor discipline within the big picture of trying to figure out what processes were responsible for creating the solid stuff that makes up the earth. Not only that, but it is the study of a catastrophic process and many geologists have trouble figuring out just how important such events are in the sediment history of our world. In fact, the interpretation of sediments and how they got there has long been the preserve of a more stabilist (stabilism is the idea that the earth changes very slowly over time) approach. As the great Bill Dickinson stated in his 1971 paper in the journal Science:

> Geologists have long recognised that drastic changes in regional geologic features and organic evolution in the past are recorded by rocks and fossils. Despite these ruling concepts of change as the order of the world, many geologists until recently had a stabilist view of global geographic features in earth history . . . During the past decade the emergence of a mobilist view, here called the plate tectonic theory, has changed the outlook . . . Acceptance of the mobilist view changes many ideas about geologic history.

That was in 1971 and while many new mobilist concepts (mobilism allows for relatively rapid changes in Earth's history) such as tsunami geology have come out of the woodwork since then, they are in many ways still viewed with some degree of skepticism. After all as many a geologist will cry, can you really identify tsunami deposits that are 2.5 million years old? Surely they would have been destroyed by now? I find this a somewhat amusing attitude given that the same premise is never truly applied to the more prosaic and ongoing processes that apparently dominate the sedimentary world. Hmmmm—which is more likely to be preserved, the regular, ongoing processes such as the gentle low-energy deposition of fine sediments by rivers as they approach the sea and form large estuaries or sudden, rare, high-energy processes such as tsunamis, rampaging inland, reshaping the coast, and laying down sediments in places where they are much less likely to be eroded?

Now that I have waved a flag here in favor of mobilists—or dynamicists of whatever we call ourselves—I can say that there must be balance. It is not all Hollywood action thrillers in the sediment world. Even within the tsunamis world, we must take care when it comes to the really big stuff—whale bones on top of cliffs, boulder deposits strewn along coastlines and clifftops, and so on. To propose a tsunami origin for these requires far more careful consideration than simply saying, "Well, that's unusual, must have been put there by a tsunami."

With this in mind, the identification of any sediments that might be related to the Eltanin tsunami must be well and truly put through the wringer—almost any other interpretation is probably likely to be a more reasonable one. Hornitos—nah, it doesn't really make the grade—after all, the deposits may be quite high above sea level, but then thinking about the sea level 2.5 million years ago, how much the coastline has changed since then, what other activity was going on in the area at the time, etc . . . . To cut a long story short, a 2014 paper has a blunt argument for "Hornitos as a debris flow deposit caused by an earthquake . . . ." In other words, there was no tsunami involved other than the possibility that this large debris flow or landslide off what were then coastal cliffs may itself have generated one. Eltanin had nothing to do with it, unless of course the massive impact of the asteroid was the cause of the earthquake! Hmmmm.

Hornitos now slips into geological obscurity, which is not such a bad thing. It is not the most wonderful place to visit to be honest, although it has a gorgeous sandy beach that is both its one saving grace and its ultimate downfall. The town is mainly comprised of holiday homes for people from Antofagasta, and what with that and the inevitable influx of Chilean tourists to enjoy this lovely sandy beach, the environment is suffering from this seasonal overload. I am speaking from the perspective of both a geologist and a foreign visitor. The low cliffs rising on the landward side of the beach are dotted with caves and quiet little nooks and crannies that would be wonderful to explore and as someplace to set up a little seaside family enclave to enjoy a day at the beach. If I was in any other part of the world, I could see myself stopping for a while here to relax and smell the roses so to speak, but not here. Every nook, every cranny, every cave defies the desire to stop and smell the roses because the smell is anything but pleasant, almost the entire expanse of picturesque coastal strip is littered with feces and used toilet paper and dominated by the overwhelming stench of human sewage.

We move on.

It is all too easy to argue about sedimentary evidence that is 2.5 million years old. After all much of what we would like to be able to use is well past its sell by date and a lot of the detail has been lost. Is a debris flow a tsunami deposit or something else? Hard to ignore though, are the bones . . . these do leave something of a conundrum lying there in the ground.

These bone-beds are an unusual mixture of complete and fragmented marine and terrestrial mammal skeletons. The majority of the fauna represented by the skeletons indicate calm water environments, and yet the excellent preservation of even extremely fragile bones suggests rapid burial. The sediments encasing them consist of a wide variety of materials bearing many hallmarks of rapid and catastrophic deposition—rip-up clasts, pebbles, and fining-upward sediments. This juxtaposition of rapid and catastrophic emplacement of sediments coupled with the preservation of fragile skeletal remains is rarely observed in storm deposits, but is common in tsunami sediments. Could these be an indication of the Eltanin? The jury is still out on this one. Nothing is ever simple when dealing with such old sediments, and they may well have been deposited over several events. But what events? At least the Eltanin tsunami is not yet dead and buried, and perhaps the answer to the nature and extent of this event will finally be solved in bones, not in sediment. The value of such evidence may lie in its indication of the effects of catastrophic tsunamis on life on this planet. Yet few tsunami researchers pay much attention to these details, largely because they fall outside their expertise. Those who do have such expertise rarely if ever consider tsunamis, because tsunamis fall in turn outside their expertise.

We do science itself an injustice when we stuff it into siloes, but that doesn't stop us doing it.

With evidence for the Eltanin tsunami still proving elusive, there was little to do but move on and use the opportunity of a trip to northern Chile to try to finally find some more recent, and potentially unique, tsunami deposits that could speak to us about the more recent past. In such a harsh and brutal environment, finding the Mother Lode of multiple tsunami deposits in one place preserved in the Atacama Desert was at best a long shot.

Immediately to the south of Hornitos is the Mejillones Peninsula. Ah ha, this is the earthquake sticking point, the place that seems to get in the way of a really, really big earthquake happening along that coastline. Big ones to the north and big ones to the south, but nothing spanning this lump of land that sticks out into the sea just north of Antofagasta. To most researchers this represents a full stop, or a geological page break, that divides the stories of earthquakes in the region. However, recent work is starting to question this wisdom. Wouldn't it be nice to find some spot on the peninsula that might have a record of past tsunamis, one perhaps that mixes evidence from the north with evidence from the south, perhaps even evidence of some events that spanned the divide?

As fieldtrips go the search for Eltanin, and for more recent events, had been a grueling road trip. Hundreds of kilometers of road travel, baking hot days digging under the never-ending sun, and a final day stymied by time limitations that made it impossible for us to get to the site we wanted to visit on the northern side of Mejillones Peninsula. Sitting by the field vehicle, starting to accept the inevitability of failing at this final hurdle, we stared to the south. Isla Santa Maria lay just offshore, basking in the stark beauty of the azure blue seas. Just to the south of it there seemed to be an inlet, a funnel-shaped bay facing northwest, at the head of which was a beach popular with weekend campers—Caleta Errazuriz.

One last roll of the dice?

It was short drive, a matter of a few kilometers. We drove down the coast, eased our way off the gravel road, and drove out onto the flat surface immediately landward of a small beach ridge. If this had been any other part of the world, I would have expected to see a nice little wetland, but here it was just fine sand and as dry as a bone. Hmmm—but if it operated just like a wetland in that it was a quiet sleepy backwater that could get wet from time to time, then surely any large catastrophic event would come rampaging in, deposit its sediment, and leave it there. What if???

As field sites go this must be one of the easiest places that I have ever worked. We stopped in the middle of this pseudo-wetland, jumped out spades in hand and started digging—literally just a couple of steps from the vehicle. Amazingly, within a matter of minutes we had dug a trench a meter deep and were transfixed by what we had found. This was a layer cake to end all layer cakes. The fine sands that dominated this flat expanse of coastline were inter-fingered with coarse gravel layers replete with marine shells. In our briefest of brief visits, a mere hour or so, we counted, sampled, and recorded NINE layers (Figure 9.4).

And that was it. No more, no time. The most incredible, unexpected find of the entire trip lay before us and we had to leave. No more funding, no planned return trips, just the memories and photographs of what was potentially a stunning site. Were they tsunami deposits? Were they storm deposits? How old were they? We had found our El Dorado and had to walk away.

Is there a happy ending here? I must admit to a little professional secrecy here. In my field notes I named the site Bahia sin Nombre—the bay with no name—and in all honesty I did not know its name at the time. And so it remained. Unknown, unvisited, awaiting funds that would doubtless never be forthcoming as is the way of these things.

Figure 9.4 Caleta Errazuriz (I named it Bahia sin Nombre)—Multiple layers of marine derived material sitting between terrestrial sediments from the surrounding hills, with me for scale. (Image: J. Goff)

Five years later, archaeology came calling in the form of that serendipitous email. A group of Chilean archaeologists had found some cool stuff all along the coast that I had visited, was I interested?

Too bloody right I was.

10

Strand 9

Life and Death on the Edge

> However, due to the stability, richness and predictability of marine
> resources, there is a continuous cultural history of 13,000 years in
> the area, showing an efficient adaptation and traditional ways of life
> persisting for several millennia.
>
> S. Rebolledo et al. Maritime fishing during the Middle Holocene
> in the hyperarid coast of the Atacama Desert

Archaeological surveys over many years along the Atacama Desert coastline
have started to piece together the chronology of human existence there. The
dominant lifestyle was, not surprisingly, some form of hunting and gathering, with the community making use of the slim pickings along the coast
and the amazingly rich and diverse food sources lurking just offshore.

As with all such studies, trends can be picked up over the millennia. These
have allowed scientists to divide the history of hunter-gatherer groups into
different periods—based on the resources they ate, the tools they used, what
they lived in, and so on. Along this coast, human history starts in the period called Archaic I, which covers a time period of some 2000 years, starting
around about 12,000 years ago. Since this represents the starter population in
the area, things were fairly ad hoc. This was the time of the Huentelauquen
Complex, a fancy phrase used to describe a population of highly mobile
hunter-gatherers that generally lived in the southern part of the region
around Antofagasta—not far from the Mejillones Peninsula. Here things
were not quite as bad as they are today, the current desert conditions not really kicking in until about 10,000 years ago. But that is not to say it was much
better, and so we see a subsistence lifestyle that revolved around eating seals,
fish, shellfish, and the odd land animal. This was a high-energy lifestyle, with
people having to move around to find what they wanted.

In Search of Ancient Tsunamis. James Goff, Oxford University Press. © Oxford University Press 2023.
DOI: 10.1093/oso/9780197675984.003.0010

For some obscure reason, around 10,000 years ago this area was abandoned for almost 1500 years. Why? Well, it is always tempting for the likes
of me to try and invoke something catastrophic, but to be honest, no idea,
although perhaps the increasingly severe desert-like conditions meant that
things were easier elsewhere. Whatever the cause, we see the place being
reoccupied at the start of what is known as the Archaic II period. While the
lifestyle was almost identical, the old camp sites used by the earlier lot were
not used this time. Now I know that there is a lot of real estate in the area, but
it seems odd that these older sites were not reoccupied because they would
presumably have been the best places to stay. This suggests to me that the
changes in climate made new demands on these people and so they sought
out different camp sites to fit their needs. Still no tsunamis though. OK, sea
level was still rising to get to present day levels but on this coast that didn't
make a huge amount of difference with regards to distance from the sea. The
real issue when it comes around to tsunamis here is that, well, no one has
looked for anything that old (apart from Eltanin) and quite frankly it would
have had to have been a real biggie for any evidence to have survived, especially since the sea level was lower at that time, so presumably there hadn't
been any asteroid dropping in (Chile's known paleotsunamis are shown in
Figure 10.1).

Life must have been good though because when we hit the Archaic III period around 7500 years ago the population was on the rise. Yes, the lifestyle
was still subsistence, but it would seem that humans were finally coming to
terms with living life on the edge. Settlements got bigger—we can see that
from the size of their rubbish dumps if nothing else. It may have been the
growing sophistication of their tools that allowed for this growth—fishhooks,
harpoons, and knives became *de rigueur*. But they also started to trade with
others and with that came more things to eat and the ability to make better
tools and plunder the seas even more. Quite naturally these communities
started to become more stable, they could get what they wanted from the sea
and so were anchored more along the littoral divide. By the time the Archaic
IV period kicks off around 5500 years ago they are building stone structures
and establishing something of a more developed society. Life was good.

And then we hit Archaic V somewhere between 4500 and 3500 years
ago. Gone are increasingly sophisticated coastal communities, replaced
by short-term occupations—essentially back to the old days. It is tough
for archaeologists to find any sense of continuity or stability. However, fish
were still the order of the day, you could easily be a pescatarian here. All in

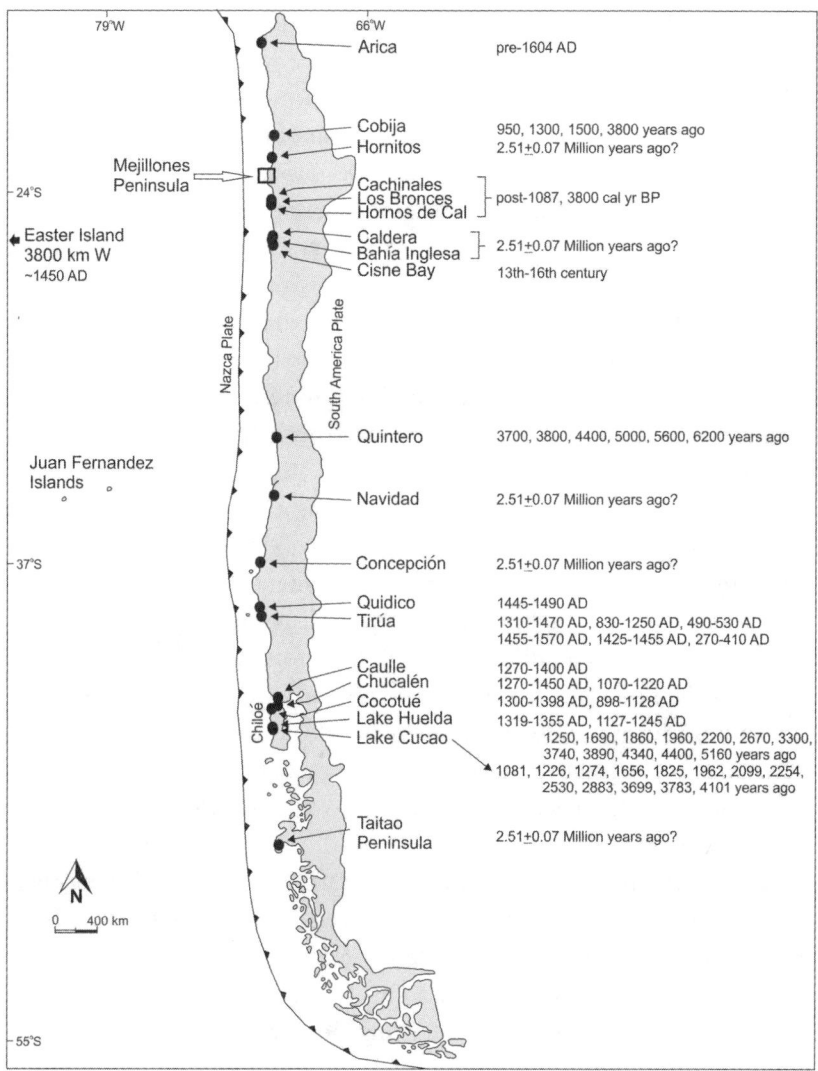

Figure 10.1 Palaeotsunamis in Chile—A not quite comprehensive list of approximate ages, since new work is coming off the presses every year. With the exception of possible Eltanin sites, in the middle and southern part of Chile events date back to a little over 6000 years (the easier place to work). To the north, a far less complete record dates back to around 3800 years ago, testament to the difficulties of working there. The big sticking point of the Mejillones Peninsula is shown—the 3800-year-old event is the only one so far found on both sides of this peninsula. (Image: J. Goff, after Goff et al., 2020)

all though, this was a major backward step in the development of coastal communities along the Atacama Desert coast.

Archaic V seems to represent about 1000 years of hell by the look of it—although given estimates in ages, the precision of radiocarbon dating, and the fragmentary archaeological record, it was probably some sudden event that occurred sometime around the middle of this age range that caused the backward step with the aftermath dragging on for some generations. But then again, maybe not, maybe it was a conscious decision to return to a more simple life—maybe. Archaic V was followed by what is known as the Formative Period when trading increased markedly, especially with inland sites, and everything got better again. There were fish, meat, pottery, and metal goods, but the sea was still their bread basket, dominating anything else that may have made it from more exotic locations. From then on, things just started to get better, or at least as "better" as it can be when you live on the edge of a desert. The multi-millennia long tradition of the exploitation of marine resources within such a hostile environment for human occupation defines human activity in this area right up to the present day.

Yes, life along this coastline was and is dangerous, and so where better to study the effects of tsunamis on marginal human populations. Over the generations these people undoubtedly got better at surviving the smaller events, after all this is not exactly a storm free coast, and a long familiarity with the sea probably induced a degree of caution into their lifestyles. But what about the bigger events?

Indeed, what about the bigger events? What about Archaic V, somewhere between 4500 and 3500 years ago? Why the sudden change from sophisticated coastal communities to short-term occupations and the major backward step in cultural development? This is precisely what we have seen throughout the southwest Pacific following the 15th century earthquake and region-wide tsunami—culture went backward.

Right then, a quick recap. The coastal Atacama Desert is not a great place to live except for one key thing, the offshore environment is one of the most productive marine ecosystems in the world because of the cold Humboldt Current system. As a result of the stability, richness, and predictability of this food source there has been continuous human occupation here for over 12,000 years. BUT, every now and again there are some really big offshore earthquakes because we have two tectonic plates, the Nazca and South American, that converge to create a subduction zone under the sea just offshore. These in turn have generated some large tsunamis such as the AD 1877

Iquique and AD 1868 Arica events. Unfortunately though, due to a variety of confounding factors we have not done a very good job up to now at figuring out how often these things happened in prehistory and how big they were. Until now, that is. The nexus between archaeological and geological research has now started to throw some light on just how bad things were, and can be, along this coastline. This all stems initially from geological work carried out at or adjacent to archaeological sites and in conjunction with archaeologists.

This is why I accepted the challenge/offer from my Chilean colleagues that I have repeatedly referred to at various times—to see if we could find any evidence for a possible tsunami around 4000 years ago, one that could have caused the marked cultural and settlement pattern changes of Archaic V. Ah ha—it is starting to come together at last.

So let us just see what this meant for me.

Apart from evidence for the somewhat unusual Eltanin asteroid tsunami that may or may not actually be evidence for it at all, no one had really found or even attempted to find anything in the Atacama Desert region earlier than a few hundred years ago. How do you do it? Where do you start?

This is where "new glasses," thinking outside the box, and getting under the skin of the environment were essential in trying to make any headway on this problem. Over the years I have found that for fieldwork in new sites there is nothing to beat being prepared. Of course there are the standard things you can do—looking at maps, satellite images, past research, geology, climate, geomorphology, etc., etc., etc. Everyone has their own way of getting ready for work at a new site, some just wing it, drop in, do their stuff, and get out—all very confident and swashbuckling in a way but to be honest, absolutely stupid. However good a researcher might be and however much preparation you do in the office, lab, library, or wherever, there is absolutely nothing to beat on-the-ground pondering. By this I mean that it is imperative to factor in some quality time at the start of any fieldtrip to simply sit there (stand there, walk around) and absorb the context of the site. It might be because I was originally trained as a geomorphologist—someone who studies the processes that go about creating the physical features of the surface of the earth—or it might simply be a desire to engage in some way with the site. Whatever it is, as soon as I have the opportunity, I will spend time checking out how a particular site "sits" within the landscape and its environment. What physical processes such as say waves, rivers, landslides, and so on have shaped and created what we see today and how might these have operated there in the past?

For me this is fundamental to understanding where I want to look for any evidence of past tsunamis. For example, if I want to look for tsunami deposits in a coastal wetland in New Zealand, say, where do I dig holes or take cores? I could just make the place look like Swiss Cheese and run around taking endless cores, digging a ridiculous number of trenches, and covering the entire area. More reasonably though, I could sit back and look at the place in detail and see where, for example, a river that flows into it today might have flowed 500 years ago and in doing so obliterated any earlier tsunami deposits—not a great place to look if I am trying to study a multi-millennial record of the events. The list of permutations and combinations that come up when you do this are almost endless, but quietly processing all of the details in my brain inevitably fine-tunes the plan and identifies what might best be termed "the low hanging fruit"—the best places to look. This approach may seem blindingly obvious, but it is surprising how many of my colleagues simply rely on what they have discovered in the office or what their students have found out for them (a common practice) and just charge in like a bull in a china shop. While their approach may well produce good results, it never produces results that are better than a more considered approach, and often I am left banging my head against the proverbial brick wall when I read or listen to their subsequent academic prognostications. It is often what they did not study as opposed to what they did study that is the highlight of their work for me. I have seen a seemingly endless litany of over-funded projects with half-baked results that would have been so much better if they had simply taken the time to think about what they were doing. That is not to say that all of my work smells of roses. Just look at my track record in the Chatham Islands for example. As the old saying goes "there is no one more virtuous that a reformed whore." The point is, you reform.

For northern Chile, the start of the campaign was not an auspicious one. At the time of this work I lived in Australia and so the flight was not too bad, about 14 hours, plus the time difference, about 14 hours, and then the extra flight up to Antofagasta, about two hours, and then the car journey south of Antofagasta to Taltal—about three hours or so. Taltal is still very much in the Atacama Desert and continues our journey south down the coast that we started on in the last chapter. It is about as far as we need to go right now. Oh yes, and then dinner, drinks, and bed at about 2:00 in the morning, Chilean time.

The next day started at 8:00 a.m. Not surprisingly, I was on the tired side of awake. However, as always the tried and tested fieldwork adage applied—"work hard, play hard"—so it really didn't matter how tired I was. On the way to Taltal we had spent a considerable amount of time driving the inland Ruta 5 that extends the length of Chile as far south as the town of Quellón on the island of Chiloé some 29 hours or 2600+ km (1600+ miles) away by car. But the final 40 minutes of our journey clung to the narrow coastal strip between Paposa and Taltal. At its widest this narrow, moderately flat, coastal strip was maybe a kilometer wide, at its narrowest the road had somehow been hacked out of the cliffs, perched precariously above the sea, and backed by a decidedly unstable rockface that needed only a minor earthquake to help it fall down. Apparently it has done so many times.

There is nothing quite like perspective, and I was getting plenty of it. This was a harsh place, made even more so by the geomorphology. OK, so it was a desert, hardly any vegetation, and so on. But to be confined to such a narrow coastal strip exposed communities to the vagaries of the sea (El Niño driven storms, tsunamis), earthquakes (landslides), and intense El Niño driven rainfall events. These last cause extreme flooding as waters rampage down the almost endless number of *quebradas* or ravines that brutally cut through the steep cliffs behind. The *quebradas* deliver huge quantities of water and sediment down from the hills, inundating the coastal strip, blocking routeways, either eroding everything in their path or smothering it in thick layers of sand, gravel, and boulders. A rough place to live, but to my more immediate concern, a tough place to do work.

Where do you look for tsunami sediments? There was little accommodation space—that zone between the coast and the full extent of the tsunamis runup where sediments could be deposited—and what there was of it didn't look entirely inviting. Unstable slopes prone to landslides—nothing there, bare narrow strips of sandy coast exposed to strong winds and the sea, a complete lack of wetlands, a complete lack of lakes, no river estuaries, just desert. Hmmm—tricky.

Over the ubiquitous early morning instant coffee (sometimes you just have to bite the bullet—ground coffee is as rare as hen's teeth in rural Chile) everyone was eager to get out and start looking for tsunami deposits. I could see that as something like a search for the Holy Grail, an impossible quest for a mythical object. As coastal environments go, this was about as bad as it gets. Yes, we had success at a few sites to the north, and my thoughts turned lovingly to Caleta Errazuriz. Oh to have something like that, which would indeed be the Holy Grail, but that was not to be—nothing so simple. I must admit to feeling pretty nervous about this search. It seemed incredibly unlikely that any tsunami deposit could withstand the rigors of this climate and geomorphology for a year let alone thousands of years. While I had had a modicum of success to the north, this was a different proposition altogether.

We had 10 days in the field, and I made the not particularly popular decision that the only way to make any of those days meaningful was to spend the first two of them simply getting to know the environment. I needed a car, someone who could speak English (to this day I still cannot speak Spanish and it irks me beyond belief, the words drift in one ear and out of the other and never even bother to try and stick around on the way through, but I keep trying. For some bizarre reason though I seem unable to forget the word "estacionamiento" or "parking lot," which is not particularly useful if I am honest), and I needed to cover a lot of coastline. It came as a bit of a surprise to me that I suggested this. I did not have that almighty confidence that most researchers have when they swagger on to a site and immediately do their stuff. Having said that, I could not think of any colleague of mine who had ever had such a daunting task, so I wondered whether they would even have considered trying. This was very much outside your average tsunami researcher's comfort zone.

And so it began.

The main road was meanderingly frustrating, mostly staying away from the bits I wanted to get at, but then to the student who was helping me this happily appeared to be no problem. Not only that but he was aiming to do a thesis on the topic of tsunamis in the area. No pressure then—find the impossible so that he would have a thesis to write . . . I should say that he is now in Australia doing a PhD on tsunamis (something that would have been impossible in my student days since tsunami geology hadn't been invented then), and that is very pleasing. But those days were ahead of us. There were plenty of what might be politely called minor "gravel" roads that were sort of road-like in the sense that they represented ribbons of moderately distinct routeways through the unremitting veneer of gravel and sand that comprised the land around us. Some seemed impossibly steep, others decided to not be roads at some points, and there were potholes aplenty. In other words—perfect stuff for good rugged fieldwork.

I didn't dig a single hole in those two days—no spade work, no trowel scrapings, just looking and pondering. The landscape was a fascinating melange of the kind of hell that could only be inflicted upon a coastline by a combination of nearly all the extreme processes you care to mention. But a pattern emerged. The rocky coastline was interspersed with little gravel pocket beaches and the occasional vast flat expanse of sand, but in many places the mountains came all the way down to the sea and where they didn't there was almost always the inevitable *quebrada* that while dry showed evidence a-plenty of the vast power of the floodwaters that would on occasion belch forth. They were choked with huge quantities of gravel and sand that spilled out as giant fans onto any flat land available—this was geomorphology on steroids. The sea was littered with bedrock pinnacles rising out of the water, but these were also matched by similar pinnacles now stranded well above high water, isolated some 5 m (16 ft) or so above the tide. Ah ha, we had an entire uplifted seafloor that had been raised up by at least one, maybe more, large earthquakes sometime in the past. These stranded pinnacles stood proud of the gravel fans creating a weird landscape looking remarkably like a miniature mountain range perched between its big brother inland and the sea.

The two days were complete, as was my mental map of the coastline. I knew where to look, or rather, I knew of the only possible place where there evidence might have survived. To me it seemed to be a long shot, and so I played my cards close to my chest. If it didn't work, I would have to resort to the tried and tested technique of making the entire place look like Swiss Cheese on the off chance of finding something—not a very edifying prospect.

Day three and it was time to put my theory into practice, and my colleagues were along for the ride,—both geologists and archaeologists—and some of their students. Joy.

My theory was simple. It could be expected that almost every time there was a decently brutal El Niño event, storms would pound the coast, so no point looking for any tsunami evidence within the storm hammered widths of the coastline. Equally, heavy rainfalls would give rise to huge flash floods that would cascade down the *quebradas*, bringing down huge amounts of water and sediment. The water would undoubtedly erode a lot of coastal sediments put there since the last floods, and the sediment would be deposited in thick layers over anything else that remained, making it impossible to find tsunami deposits whether they were there or not. That would have been like searching for the proverbial needle in a haystack—no point looking around those areas either. Add to all of this the fact that when there was no El Niño operating then winds would blow away finer sediments, at the very least, destroying the integrity of any tsunami deposit that might have been deposited. There were none of the nice quiet backwaters that would usually provide a convenient resting place for sediments, no wetlands, no lakes, no caves. BUT, there was an old coastline that had been uplifted by one or more earthquakes in the past, many meters above the present day coastline. A big tsunami could inundate this zone—while it would be unlikely that sediments carried by storms could reach it—so there might be something left behind. However, on the down side, much of this surface was exposed to the devastating effects of the *quebradas*. But that was the point though, "much" of it was, but not all of it. I theorized that there would be sites on this uplifted terrace (or terraces) that were protected from both the vagaries of storms and the *quebradas*. These sites were surrounded by rocky outcrops on all sides that would serve to protect any deposits (tsunami, archaeological) from erosion caused by active terrestrial and coastal processes. They were the sleepy backwaters in this battered landscape. In a sense they were the dry, rocky coast equivalents of wetlands.

The proof of the pudding would be in the eating, so to speak. I had identified potential "low hanging fruit," and now it was time to pick one off. I could see no obvious favorite and so off we went to dig one of them. The Los Bronces site is to the south of Taltal, where there is a narrow coastal plain. Not surprisingly it sits close to some *quebradas*, but is raised up a good 8–9 m (26–29 ft) above sea level and is well over 100 m (300+ ft) from the coast. There is also an archaeological site not far away, and so if we found any evidence for a large tsunami here it would (assuming the ages of both sites coincided) have impacted the humans living there. In many ways this was the perfect site to start with. The trench we dug ended up being almost 2 m (6.5 ft) deep but it didn't take long to find the mother lode. A little over 20 cm (8 inches) down, we hit our first tsunami deposit (Figure 10.2). Yes!

While I was sweating and panting from the efforts of digging a trench in the dry, baking heat, I felt a raft of emotions when I realized what we had found. Stunned to find it, relieved that we actually had something to look at, extraordinarily pleased that the past two days had been worthwhile and that essentially the first time we put a spade in the ground we had found something. The years of studying these beasts in a seemingly endless array of environments had paid off. Yes, the ego felt suitably massaged, but over and above that was the satisfaction that the system works. We can identify tsunami deposits even in the harshest of environments, we just have to think about it a bit.

The remaining days were filled with field research. While my gut told me that we had actually found at least two prehistoric tsunami deposits, science tends to need a bit more than that. We threw the book at them—they had survived surprisingly well.

Figure 10.2 Upper tsunami deposit at Los Bronces, full of marine sediments and abundant shell material. (Image: J. Goff)

We had sharp erosional contacts at the base where the high energy waves had eroded what was beneath. Nicely rounded boulders and gravel from the sea mixed with sand and large quantities of marine shells also told us that these had probably been quite big tsunamis. The geochemistry screamed marine origin, and at some places we actually found flame structures where the sediment had been "bent" in the direction of water flow. Above and below these deposits were angular sediments that had come out of the hills or been reworked by wind. Even these sites were not completely free from the hell of El Niño floods, but they were high enough up above main channels that burst forth out of the *quebradas* to avoid most of their wrath.

After much laboratory analysis and the important radiocarbon dating of assorted material we had two clear set of dates. One event was younger than about a 1000 years ago. It was also a far smaller event than its predecessor, which cropped up at far more places along a much longer stretch of coastline around Taltal, and that one was . . . around 3800 years old.

This was the start. Over the next few years we found evidence for this roughly 3800-year-old event impacting more than 1000 km (620 miles) of the Atacama coast and the local populations of hunter-gatherer-fishers.

The tsunami had been caused by a massive earthquake and fault rupture encompassing bits of the subduction zone that extended *across* the Mejillones Peninsula segment, the place that until then had been thought of as a major sticking point for any earthquake—it was not meant to cross that. Historical earthquakes had happened to the south and to the north of it, but they had never joined up. This really was a big one then. Some fancy computer models put this earthquake at around a Magnitude 9.5, equal to the largest ever in recorded history, which had occurred in southern Chile and had been responsible for a Pacific-wide tsunami that killed people in the Hawaiian Islands and Japan.

Archaeological evidence for the tsunami along the Atacama coast was plentiful, including partial or total destruction of structures and erosion of old middens and occupation sites. All of these sites were almost entirely abandoned following the event. The human response was clear. Stone buildings were no longer built, the populations became more mobile again, never staying in one place too long, but they still ate food from the sea—it was their only option. It was not until nearly 1000 years later that people started to return to the old ways, but even then most human occupation sites were located farther from the shoreline and at higher elevations. Move inland and uphill—the best response to any tsunami and a sign of strategies developed to deal with the challenges of such episodic catastrophic events. It also appears that, as is the way with humans, memories faded, and so millennia after the event communities marched closer to the sea again. Some of the long abandoned sites were even reoccupied.

As it stands this is the largest earthquake and tsunami that has ever been found in Northern Chile. It had a catastrophic effect on the local populations and the tsunami rampaged across the Pacific Ocean to smash into the Chatham Islands in New Zealand (Chapter 5).

There is a more modern parallel here, that of the AD 1700 Cascadia earthquake and tsunami. It popped up in Chapter 1 and quite remarkably was dated to around 9:00 p.m. Pacific standard time on January 26, AD 1700. While the dating and detail reported for this event benefited greatly from historical Japanese accounts and a wealth of data from wetlands and ghost forests in the US Pacific Northwest, it was nonetheless an amazing piece of scientific research. How does the 3800-year-old event stack up?

It occurred in prehistory, no convenient written records here, and while the AD 1700 event was prehistoric in the northwestern United States it was still recorded in the oral traditions of Native Americans. The physical evidence for the 3800-year-old event sits in two of the toughest regions of the world in which to carry out tsunami research. To the east is Northern Chile—the baking hot desert coastline is almost devoid of vegetation, lacking wetlands, lakes, and forests, and exposed to the brutal effects of El Niño driven storms and floods. To the west are the Chatham Islands, almost devoid of any sediments at all, because there is nowhere for it to come from, a land where a few centimeters of sediment represents thousands of years of history, and where there was no human population until the last few hundred years. This was an event too old for oral traditions in Chile but with a fragmented, archaeological past that has amazingly revealed so much.

How does it stack up? To my somewhat biased perspective, it stands up as the best example of what can be achieved in the absence of historical information. It is precisely how prehistoric tsunamis should be studied using the full range of skillsets available, and it has the potential to go one further and be the best example of all, historical or otherwise, and that is because of prehistoric coastal mass burial sites. After all, tsunamis do not upset entire communities without a massive death toll, which demands rapid burial. We have crossed paths with these mass burial sites earlier on and made the link with tsunamis largely because of similar dates, but also because of the nature of the burials. We have been amazed at the fact that no archaeologists have really ever seriously considered a possible tsunami cause for such burials.

Well, times they are a-changing.

As part of an incredibly comprehensive archaeological review for northern Chile in particular but the Chilean coast as a whole, there is now a good understanding of the ages of most of their known coastal mass burial sites. There are many, of many ages, but of particular interest to this ancient event are sites that date to around the same time period. Examples include sites as far south as La Serena and Coquimbo, some 450 or more km (280 miles) south of Taltal. The Punta Teatinos (La Serena) site contains 198 skeletons comprising 59 females, 64 males, 10 adults of unknown sex, and 65 juvenile individuals of indeterminate sex and dates to somewhere around 4000–4900 years ago, which is about right. At El Cerrito, on the other hand, 105 individuals were found, including 27 females, 25 males, 28 adults of unknown sex, and 25 juveniles, dating to around 3800 years ago, which is even better.

And there are more sites too.

11

Strand 10

The Future

In the absence of any form of written record, determining whether a person drowned in salt- or fresh-water, is a matter for forensic archaeology. Most commonly, forensic archaeology involves the application of archaeological techniques to the search for and recovery of evidential material from crime scenes.

P. A. Andrade et al. Reconstrucción del modo de vida de individuos del arcaico de la costa arreica del norte de Chile

Mass burial sites that date to around about the same age as this massive Chilean earthquake and tsunami are all very well, but are merely interesting circumstantial evidence. Or perhaps not. Presumably, there must be a lot of other burial sites where the ongoing attrition of human life and death is charted by the gradual addition of more and more burials over time. But the mere fact that there are mass burial sites close to the coast surely speaks of a singular event that was catastrophic to the population at that time. Careful archaeological reexamination of the characteristics of these burials in the context of this new-found seismic event may cast a new light on past interpretations, or may even at last offer up a tangible reason for them being there (Figure 11.1).

But there is always more.

The search for ancient tsunamis has grown far beyond my simple efforts rummaging around in the geological ripples of these catastrophes that are preserved in the ground. Indeed, the search is now being approached from many standpoints, which is exactly as it should be and is to be encouraged. Anthropologists stalk the elusive records preserved in oral traditions, tsunami archaeologists track down the memories of past settlements, geochemists and ecologists may look in the sediment but they don't even need

In Search of Ancient Tsunamis. James Goff, Oxford University Press. © Oxford University Press 2023.
DOI: 10.1093/oso/9780197675984.003.0011

Figure 11.1 Coastal mass burial in a rockshelter at San Lorenzo, near Taltal, in northern Chile, dating to somewhere around 2000–2700 years ago—quite an age range. Currently though, there are no known ancient Chilean tsunamis within this range, although there are two or three potential candidates from across the Pacific Ocean. The absence of an associated tsunami deposit may be due to a lack of looking, a lack of preservation, or simply the fact that this is not tsunami-related at all. I feel perhaps that these multiple working hypotheses might get the approval of Thomas Chrowder Chamberlin. They key now is to see if we can use archaeology to crack this case. (Image: P. Andrade; with permission)

to be able to see any deposit to be able to find evidence, microscopic signals are the domain of a variety of people—geneticists, micro-paleontologists, radiologists—and geomorphologists can look down from on high and pick out tell-tale signs.

Every single one of these different approaches represents a strand or perhaps more than one, and every single one of these strands is reaching out to find buddies to enrich their discoveries. In almost all cases those fellow strands will come from the already tried and tested toolbox of ideas that has gradually been assembled over the past few decades of research. However, we are now seeing many new strands starting off, and not just from geology.

There is a growing inevitability that some really exciting new ideas are going to come out of the woodwork—ideas that are so completely out of left field for a geologist, for example, that they would never have seen the light of day if the study of tsunamis had remained with its head buried in the sand.

For many of us this means that we stand at a crossroads. Ahead of us is safety, the comfort of what we do, a well-trodden path that will become even more well-trodden, a modern highway countersinking itself even more into the geological landscape. While such a route will undoubtedly continue to move forward, the science will become increasingly trite and about as stimulating as watching paint dry. To the left and to the right are not just single pathways, but many. Potential innovations and excitement lie outside our comfort zone, but it is here that we can really push forward.

My part in all of this? For what it is worth, I will continue to work on tsunamis with one fundamental goal—saving lives. It is to this human side of ancient tsunamis that I am drawn, the one that in prehistory is largely the domain of archaeologists. Here there is a rich vein of unexplored avenues, or at least avenues unexplored from a tsunami point of view. This is an area where I have just enough knowledge to be dangerous, not that of an expert—I have stepped and continue to step out of my comfort zone. I come bearing gifts from across the academic divide to share and seeking to learn from others. This is where I feel that eureka moments are to be found, a lightbulb being turned on and giant leap forward being made almost in an instant.

It is here that I turn to forensic archaeology, the specialist application of archaeological techniques to the search and recovery of evidential material from crime scenes. This should not come as a surprise to anyone because, it is here we enter the realm of CSI and let's face it tsunamis are if nothing else one of nature's big killers. Are mass burials crime scenes? You bet.

Others will take other pathways at the crossroads, but just like braids they will doubtless all come together in the end. This is good because it can only mean one thing—that we are all heading in the same direction in our search for ancient tsunamis.

The one certainty in all of this work is that we will find larger events in prehistory than we have experienced in historic time. So large that when—not if—they happen again, they will be so catastrophic that we, scientists and public alike, will be caught completely off guard—unless we do some work now. The clock is ticking. We already know that events such as the AD 1700 Cascadia tsunami will be repeated. The impact of such an event on coastal communities will be so much worse the next time. What about the next asteroid strike? What about the next undersea volcanic eruption? It is all too easy to sit back and think "it will never happen to me," travel to exotic countries for holidays, bask on sun-drenched beaches, and never give a thought to the unseen planetary clock that is ticking, counting down to its next surprise attack.

Back to work (Plate 11.1)... watch this space.

Bibliography

Preface

Hall, J. (1815) IV. On the revolutions of the Earth's surface. *Earth and Environmental Science Transactions of the Royal Society of Edinburgh*, 7, 139–167.

Chapter 1. Serendipity—An Introduction

Atwater, B.F., Musumi-Rokkaku, S., Satake, K., Tsuji, Y., Ueda, K., and Yamaguchi, D.K. (2016) *The orphan tsunami of 1700: Japanese clues to a parent earthquake in North America*. University of Washington Press, USA.

Glikson, A.Y. (2004) Early Precambrian asteroid impact-triggered tsunami: Excavated seabed, debris flows, exotic boulders, and turbulence features associated with 3.47–2.47 Ga-old asteroid impact fallout units, Pilbara Craton, Western Australia. *Astrobiology*, 4, 19–50.

Goff, J., and Chagué-Goff, C. (2009) Cetaceans and tsunamis—whatever remains, however improbable, must be the truth? *Natural Hazards and Earth System Sciences*, 9, 855–857.

Goff, J., and Dudley, W. (2021) *Tsunami: The world's greatest waves*. Oxford University Press, USA.

Liu, P., L.-F., Lynett, P., Fernando, J., Jaffe, B.E., Fritz, H., Higman, B., Morton, R., Goff, J., and Synolakis, C. (2005) Learning from earthquakes: The great Sumatra earthquake and Indian Ocean tsunami of December 26, 2004. *EERI Special Earthquake Report*, 1–4.

McMillan, A.D., and Hutchinson, I. (2002) When the mountain dwarfs danced: Aboriginal traditions of paleoseismic events along the Cascadia Subduction Zone of western North America. *Ethnohistory*, 49, 41–68.

McLeod, H.M. (1913) Pre-Pākehā occupation of Wellington District. *The Early Settlers and Historical Association of Wellington*, 1, 114–117.

Power, W. (comp.) (2013) Review of tsunami hazard in New Zealand (2013 update). GNS Science Consultancy Report 2013/131.

Range, M.M., Arbic, B.K., Johnson, B.C., Moore, T.C., Titov, V., Adcroft, A.J., Ansong, J.K., Hollis, C.J., Ritsema, J., Scotese, C.R., and Wang, H. (2022) The Chicxulub impact produced a powerful global tsunami. *AGU Advances*, 3, e2021AV000627.

Spiske, M., Piepenbreier, J., Benavente, C., and Bahlburg, H. (2013) Preservation potential of tsunami deposits on arid siliciclastic coasts. *Earth-Science Reviews*, 126, 58–73.

Chapter 2. Strand 1: December 26, 2004—Indian Ocean

Dr Abalone (2014) The giant 200-foot wave at Trinidad. California. https://briantissot. com/2014/12/31/the-giant-200-foot-wave-at-trinidad-california/. Accessed July 27, 2021.

Goff, J., Chagué-Goff, C., and Nichol, S. (2001) Palaeotsunami deposits: A New Zealand perspective. *Sedimentary Geology*, 143, 1–6.

Goff, J., Chagué-Goff, C., Nichol, S.L., Jaffe, B., and Dominey-Howes, D. (2012) Progress in palaeotsunami research. *Sedimentary Geology*, 243–244, 70–88.

Goff, J., and Dudley, W. (2021) *Tsunami: The world's greatest waves*. Oxford University Press, USA.

Goto, K., Chagué-Goff, C., Fujino, S., Goff, J., Jaffe, B., Nishimura, Y., Richmond, B., Suguwara, D., Szczuciński, W., Tappin, D.R., Witter, R., and Yulianto, E. (2011) New insights into tsunami risk from the 2011 Tohoku-oki event. *Marine Geology*, 290, 46–50.

Goto, K., Hashimoto, K., Sugawara, D., Yanagisawa, H., and Abe, T. (2014) Spatial thickness variability of the 2011 Tohoku-oki tsunami deposits along the coastline of Sendai Bay. *Marine Geology*, 358, 38–48.

International Tsunami Information Centre (2004) Tsunami warning message. http://itic.iocunesco.org/index.php?option=com_content&view=article&id=1136&Itemid=2153.

Liu, P., L.-F., Lynett, P., Fernando, J., Jaffe, B.E., Fritz, H., Higman, B., Morton, R., Goff, J., and Synolakis, C. (2005) Observations by the international Tsunami team in Sri Lanka. *Science*, 308, 1595.

Morton, R.A., Goff, J., and Nichol, S. (2007) Impacts of the 2004 Indian Ocean tsunami on the southwest costs of Sri Lanka. *Coastal Sediments 2007*, American Society of Civil Engineers, New Orleans, 1–14.

Nichol, S.L., Goff, J., Devoy, R.J., Chagué-Goff, C., Hayward, B., and James, I. (2007) Lagoon subsidence and tsunami on the West Coast of New Zealand. *Sedimentary Geology*, 200, 248–262.

Reliefweb (2021) NOAA and the Indian Ocean Tsunami. https://reliefweb.int/report/bangladesh/noaa-and-indian-ocean-tsunami. Accessed July 27, 2021.

UNESCO IOC (2014) *International Tsunami Survey Team (ITST) post-tsunami survey field guide*. 2nd Edition, IOC Manuals and Guides No. 37.

Weiss, R. (2008) Sediment grains moved by passing tsunami waves: Tsunami deposits in deep water. *Marine Geology*, 250, 251–257.

Williams, S., Goff, J., Ah Kau, J., Sale, F., Chagué-Goff, C., and Davies, T. (2013) Geological investigation of palaeotsunamis in the Samoan islands: Interim field report and research directions. *Science of Tsunami Hazards*, 32, 156–175.

Yamada, M., Fujino, S., and Goto, K. (2014) Deposition of sediments of diverse sizes by the 2011 Tohoku-oki tsunami at Miyako City, Japan. *Marine Geology*, 358, 67–78.

Chapter 3. Strand 2: Same Time, Different Places

Emmons, G.T. (1911) Native account of the meeting between La Perouse and the Tlingit. *American Anthropologist, New Series*, 13, 294–298.

Goff, J. (in press) Chapter 17: The Stuttering Polynesian diaspora—environmental crises of the past and an uneasy future. In, Sugawara, D., and Yamada, K. (eds.) *Island civilizations: Implications for the future of the earth*. Springer Nature, Netherlands.

Goff, J., Chagué-Goff, C., Dominey-Howes, D., McAdoo, B., Cronin, S., Bonté-Grapetin, M., Nichol, S., Horrocks, M., Cisternas, M., Lamarche, G., Pelletier, B., Jaffe, B., and Dudley, W. (2011) Palaeotsunamis in the Pacific Islands. *Earth-Science Reviews*, 107, 141–146.

Goff, J., Charley, D., Haruel, C., and Bonte-Grapentin, M. (2008) Preliminary findings of the geological evidence and oral history of tsunamis in Vanuatu. SOPAC Technical Report No.416.

Goff, J., Hulme, K., and McFadgen, B.G. (2003) "Mystic Fires of Tamaatea": Attempts to creatively rewrite New Zealand's cultural and tectonic past. *Journal of the Royal Society of New Zealand*, 33(4), 1–15.

Goff, J., Lamarche, G., Pelletier, B., Chagué-Goff, C., and Strotz L. (2011) Palaeotsunami precursors to the 2009 South Pacific tsunami in the Wallis and Futuna archipelago. *Earth-Science Reviews*, 107, 91–106.

Goff, J., McFadgen, B.G., Chagué-Goff, C., and Nichol, S.L. (2012) Palaeotsunamis and their influence on Polynesian settlement. *The Holocene*, 22, 1061–1063.

King, D.N., and Goff, J. (2010) Benefitting from differences in knowledge, practice and belief: Māori oral traditions and natural hazards science. *Natural Hazards and Earth System Sciences*, 10, 1927–1940.

King, D., and Goff, J. (2006) Māori environmental knowledge in natural hazards management and mitigation. NIWA Client Report: AKL2006-055.

King, D., Goff, J., and Skipper, A. (2007) Māori environmental knowledge and natural hazards in New Zealand. *Journal of the Royal Society of New Zealand*, 37, 59–73.

Terry, J.P., Goff, J., Winspear, N., Bongolan, V.P., and Fisher, S. (2022) Tonga eruption and tsunami, January 2022—Globally the most significant opportunity to observe a major explosive submarine eruption since AD 1883 Krakatau. *Geoscience Letters*, 9(24), https://doi.org/10.1186/s40562-022-00232-z.

United States Geological Survey (2021) Historical perspective (continental drift). https://pubs.usgs.gov/gip/dynamic/historical.html. Accessed July 29, 2021.

Wegener, A. (1912) Die Herausbildung der Grossformen der Erdrinde (Kontinente und Ozeane), auf geophysikalischer Grundlage. *Petermanns Geographische Mitteilungen* (in German), 63, 185–195, 253–256, 305–309.

Wegmann, C.E. (1939) Die wissenschaftliche Gesichtspunkte (The scientific point of view). *Geologische Rundschau*, 30, 389.

Chapter 4. Strand 3: New Light through Old Windows

Anon. (1867) Shipping intelligence. Port of Oamaru. *North Otago Times*, Volume VIII, Issue 209, May 24, page 2.

Anon. (1867) News of the week. *Otago Witness*, Issue 808, May 25, page 11.

Atwater, B.F., Cisternas, M., Yulianto, E., Prendergast, A.L., Jankaew, K., Eipert, A.A., Fernando, W.I.S., Tejakusuma, I., Schiappacasse, I., and Sawai, Y. (2013) The 1960 tsunami on beach-ridge plains near Maullín, Chile: Landward descent, renewed breaches, aggraded fans, multiple predecessors. *Andean Geology*, 40, 393–418.

Brunner, T. (1952) *The Great Journey: An expedition to explore the interior of the middle island, New Zealand, 1846–1848*. Pegasus Press, Christchurch.

Chan, A., Goff, J., Chagué-Goff, C., Gadd, P., Bilham, R., Yamada, M., and James, I. (2016) Previously undocumented widespread subsidence on the western side of the Alpine Fault, New Zealand. Abstract 3020. 35th International Geological Congress, Cape Town, South Africa. August 27–September 4.

Goff, J., and Dudley, W. (2021) Chapter 7: Tsunamis and the U.S. Navy. In, Goff, J., and Dudley, W. *Tsunami: The world's greatest waves*. Oxford University Press, USA, 73–85.

Goff, J., Hicks, D.M., and Hurren, H. (2007) Tsunami geomorphology in New Zealand. National Institute of Water & Atmospheric Research Technical Report No. 128.

Goff, J., Lane, E.M., and Arnold, J. (2009) The tsunami geomorphology of coastal dunes. *Natural Hazards and Earth System Sciences*, 9, 847–854.

Goff, J., McFadgen, B.G., and Chagué-Goff, C. (2004) Sedimentary differences between the 2002 Easter storm and the 15ᵗʰ Century Okoropunga tsunami, southeastern North Island, New Zealand. *Marine Geology*, 204, 235–250.

Goff, J., McFadgen, B.G., Wells, A., and Hicks, M. (2008) Seismic signals in coastal dune systems. *Earth-Science Reviews*, 89, 73–77.

Goff, J., Wells, A., Chagué-Goff, C., Nichol, S.L., and Devoy, R.J.N. (2004) The elusive AD 1826 tsunami, South Westland, New Zealand. *New Zealand Geographer*, 60, 14–25.

King, D., and Goff, J. (2010) Benefitting from differences in knowledge, practice and belief: Māori oral traditions and natural hazards science. *Natural Hazards and Earth System Sciences*, 10, 1927–1940.

McFadgen, B.G. (1980) A stone row system at Okoropunga on the southeast Wairarapa Coast and inferences about coastal stone rows elsewhere in central New Zealand. *New Zealand Journal of Science*, 23, 189–197.

Simpkin, T., and Fiske, R.S. (1980) *Krakatau, 1883: The volcanic eruption and its effects (Krakatoa)*. Smithsonian Books, Washington, DC.

Chapter 5. Strand 4: Chatham Islands—Rekohu/Wharekauri— How Big?

Frohlich, C., Hornbach, M.J., Taylor, F.W., Shen, C.C., Moala, A., Morton, A.E., and Kruger, J. (2009) Huge erratic boulders in Tonga deposited by a prehistoric tsunami. *Geology*, 37, 131–134.

Goff, J. (2021) New Zealand's tsunami death toll rises. *Natural Hazards*, 107, 1925–1934.

Goff, J., Goto, K., Chagué, C., Watanabe, M., Gadd, P., and King, D. (2018) New Zealand's most easterly palaeotsunami deposit confirms evidence for major trans-Pacific event. *Marine Geology*, 404, 158–173.

Goff, J., Nichol, S.L., Chagué-Goff, C., Horrocks, M., McFadgen, B., and Cisternas, M. (2010) Predecessor to New Zealand's largest historic trans-South Pacific tsunami of 1868AD. *Marine Geology*, 275, 155–165.

Goto, K., Miyagi, K., Kawamata, H., and Imamura, F. (2010) Discrimination of boulders deposited by tsunamis and storm waves at Ishigaki Island, Japan. *Marine Geology*, 269, 34–45.

Hawke's Bay Herald. (1868) Chatham Islands. September 12, 3. http://paperspast.natlib.govt.nz/cgi-bin/paperspast?a=dandd=HBH18680912.2.16.

Holt, K.A., Wallace, R.C., Neall, V.E., Kohn, B.P., and Lowe, D.J. (2010) Quaternary tephra marker beds and their potential for palaeoenvironmental reconstruction on Chatham Island, east of New Zealand, southwest Pacific Ocean. *Journal of Quaternary Science*, 25, 1169–1178.

Lane, E., Arnold, J., Bind, J., Sykes, J., and Williams, S. (2016) Regional and distant source tsunami inundation modelling for Chatham Island. NIWA Client Report No 2016054CH.

Lavigne, F., Morin, J., Wassmer, P., Weller, O., Kula, T., Maea, A.V., Kelfoun, K., Mokadem, F., Paris, R., Malawani, M.N., and Faral, A. (2021) Bridging legends and science: field evidence of a large tsunami that affected the Kingdom of Tonga in the 15th century. *Frontiers in Earth Science*, 9(748755), 1–15.

Lomnitz, C. (2004) Major earthquakes of Chile: A historical survey, 1535–1960. *Seismological Research Letters*, 75, 368–378.

McFadgen, B.G. (1994) Archaeology and Holocene sand dune stratigraphy on Chatham Island. *Journal of the Royal Society of New Zealand*, 24, 17–44.

Nichol, S.L., Chagué-Goff, C., Goff, J., Horrocks, M., McFadgen, B.G., and Strotz, L. (2010) Geomorphology and accommodation space as limiting factors on tsunami deposition: Chatham Island, southwest Pacific Ocean. *Sedimentary Geology*, 229, 41–52.

Proctor, R.A. (1870) The greatest sea-wave ever known. *Fraser's Magazine*, 2, 93–99.

Thomas, K.L., Kaiser, L., Campbell, E., Johnston, D., Campbell, H., Solomon, R., Jack, H., Borrero, J., and Northern, A. (2020) Disaster memorial events for increasing awareness and preparedness: 150 years since the Arica tsunami in Aotearoa-New Zealand. *Australian Journal of Emergency Management*, 35, 71–78.

Yamada, M., Fujino, S., and Goto, K. (2014) Deposition of sediments of diverse sizes by the 2011 Tohoku-oki tsunami at Miyako City, Japan. *Marine Geology*, 358, 67–78.

Chapter 6. Strand 5: Just Part of the Problem

Chamberlin, T.C. (1890) The method of multiple working hypotheses. *Science*, 15, 92–96.

Chagué-Goff, C., Andrew, A., Szczuciński, W., Goff, J., and Nishimura, Y. (2012) Geochemical signatures up to the maximum inundation of the 2011 Tohoku-oki tsunami—implications for the 869 AD Jogan and other palaeotsunamis. *Sedimentary Geology*, 282, 65–77.

Chan, A., Goff, J., Chagué-Goff, C., Gadd, P., Bilham, R., Yamada, M., and James, I. (2016) Previously undocumented widespread subsidence on the western side of the Alpine Fault, New Zealand. Abstract 3020. *35th International Geological Congress*, Cape Town, South Africa. August 27–September 4.

Dobson, A.D. (1930) *Reminiscences of Arthur Dudley Dobson*. Whitcombe and Tombs, Wellington, NZ.

Goff, J., Goto, K., Ebina, Y., and Terry, J. (2016) Defining tsunamis: Yoda strikes back? *Earth-Science Reviews*, 159, 271–274.

Goff, J., Knight, J., Sugawara, D., and Terry, J. (2016) Anthropogenic disruption to the seismic driving of beach ridge formation: The Sendai coast, Japan. *Science of the Total Environment*, 544, 18–23.

Goff, J., and McFadgen, B.G. (2002) Seismic driving of nationwide changes in geomorphology and prehistoric settlement—a 15th century New Zealand example. *Quaternary Science Reviews*, 21, 2229–2236.

Goff, J., and Sugawara, D. (2014) Seismic driving of sand beach ridge formation in northern Honshu, Japan? *Marine Geology*, 358, 138–149.

McFadgen, B.G. (2007) *Hostile Shores: Catastrophic events in prehistoric New Zealand and their impact on Māori coastal communities*. Auckland University Press, Auckland, NZ.

McFadgen, B.G., and Goff, J. (2005) An earth systems approach to understanding the tectonic and cultural landscapes of linked marine embayments: Avon-Heathcote Estuary (Ihutai) and Lake Ellesmere (Waihora), New Zealand. *Journal of Quaternary Science*, 20, 227–237.

Moseley, M.E., Tapia, J., Satterlee, D.R., and Richardson III, J.B. (1992) Flood events, El Niño events, and tectonic events. In, Ortlieb, L., and Macharé, J. (eds.), *Paleo ENSO Records, International Symposium Extended Abstracts, ORSTOM-CONCYTEC.* Lima, 207–212.

Moseley, M.E., Wagner, D., and Richardson, J.B. (1992) Space shuttle imagery of recent catastrophic change along the arid Andean coast. In, Johnson, L.L., and Stright, M. (eds.), *Paleoshorelines and prehistory: An investigation of method.* CRC Press Inc., Boca Raton, USA, 215–235.

Saino, H. (2015) Tsunami disasters of Yayoi period and Heian period in the Sendai Plain. *Proceedings of the Symposium on Traces and Experiences of Past Tsunami Disasters in the Pacific Rim, and the Succession of Knowledge.* UN World Conference on Disaster Risk Reduction, March 14–18, Sendai, Japan.

Wartman, J., Dunham, L., Tiwari, B., and Pradel, D. (2013) Landslides in eastern Honshu induced by the 2011 Tōhoku Earthquake. *Bulletin of Seismological Society of America*, 103 (2B), 1503–1521,

Wells, A., and Goff, J. (2007) Coastal dunes in Westland, New Zealand, provide a record of paleoseismic activity on the Alpine fault. *Geology*, 35, 731–734.

Wells, A., and Goff, J. (2006) Coastal dune ridge systems as chronological markers of paleoseismic activity—a 650 year record from southwest New Zealand. *The Holocene*, 16, 543–550.

Chapter 7. Strand 6: The Human Touch

Bruins, H.J., and van der Plicht, J. (2017) The Minoan Santorini eruption and its 14C position in archaeological strata: Preliminary comparison between Ashkelon and Tell El-Dabca. *Radiocarbon*, 59, 1295–1307.

Durband, A.C., and Creel, J.A. (2011) A reanalysis of the early Holocene frontal bone from Aitape, New Guinea. *Archaeology in Oceania*, 46, 1–5.

Fenner, F.J. (1941) Fossil human skull fragments of probable Pleistocene age from Aitape, New Guinea. *Records of the South Australian Museum*, 6, 335–356.

Goff, J. (in press) Chapter 17: The Stuttering Polynesian diaspora—environmental crises of the past and an uneasy future. In, Sugawara, D. and Yamada, K. (eds.) *Island civilizations: Implications for the future of the earth.* Springer Nature, Netherlands.

Goff, J., Golitko, M., Cochrane, E., Curnoe, D., and Terrell, J. (2017) Reassessing the environmental context of the Aitape Skull—the oldest tsunami victim in the world? *Plos One*, 12, e0185248.

Goff, J., Lamarche, G., Pelletier, B., Chagué-Goff, C., and Strotz, L. (2011) Palaeotsunami precursors to the 2009 South Pacific tsunami in the Wallis and Futuna archipelago. *Earth-Science Reviews*, 107, 91–106.

Goff, J., McFadgen, B.G., and Marriner, N. (2021) Landscape archaeology—the value of context to archaeological interpretation: A case study from Waitore, New Zealand. *Geoarchaeology*, 36, 768–779.

Goff, J., Witter, R., Terry, J., and Spiske, M. (2020) Palaeotsunamis in the Sino-Pacific region. *Earth-Science Reviews*, 210, 103352.

Goodman-Tchernov, B.N., and Austin, J.A., Jr. (2015) Deterioration of Israel's Caesarea Maritima's ancient harbor linked to repeated tsunami events identified in geophysical mapping of offshore stratigraphy. *Journal of Archaeological Science: Reports*, 3, 444–454.

Goto, K., Goff, J., and Paris, R. (2021) Preface of the Special Issue in Earth-Science Reviews: Ten years since the 2011 Tohoku-oki tsunami—Progress in Paleotsunami Research. *Earth-Science Reviews*, 216, 103598.

Hoffmann, N., Master, D., and Goodman-Tchernov, B. (2018) Possible tsunami inundation identified amongst 4–5th century BCE archaeological deposits at Tel Ashkelon, Israel. *Marine Geology*, 396, 150–159.

Hossfeld, P.S. (1964) The Aitape calvarium. *Australian Journal of Science*, 27, 179.

Hossfeld, P.S. (1949) The stratigraphy of the Aitape skull and its significance. *Transactions of the Royal Society of South Australia*, 72, 201–207.

Kirch, P.V. (1982) A revision of the Anuta sequence. *Journal of the Polynesian Society*, 91, 245–254.

Maselli, V., Oppo, D., Moore, A.L., Gusman, A.R., Mtelela, C., Iacopini, D., Taviani, M., Mjema, E., Mulaya, E., Che, M., and Tomioka, A.L. (2020) A 1000-yr-old tsunami in the Indian Ocean points to greater risk for East Africa. *Geology*, 48, 808–813.

McFadgen, B.G., and Goff, J. (2007) Tsunamis in the archaeological record of New Zealand. *Sedimentary Geology*, 200, 263–274.

McSaveney, M., Goff, J., Darby, D., Goldsmith, P., Barnett, A., Elliott, S., and Nongkas, M. (2000) The 17th July 1998 Tsunami, Sissano Lagoon, Papua New Guinea—evidence and initial interpretation. *Marine Geology*, 170, 81–92.

Mjema, E. (2018) Catastrophes and deaths along Tanzania's western Indian Ocean coast during the Early Swahili period, AD 900–1100. *Azania: Archaeological Research in Africa*, 53, 135–155.

Morgan, O.W., Sribanditmongkol, P., Perera, C., Sulasmi, Y., Van Alphen, D., and Sondorp, E. (2006). Mass fatality management following the South Asian tsunami disaster: Case studies in Thailand, Indonesia, and Sri Lanka. *PLOS Medicine*, 3, e195.

Nason-Jones J. (1930) *Notes on the Geology of the Finsch Coast Area, northwest New Guinea*. Anglo Persian Oil Company Ltd, BASE Company Report APOC 26.

Reinhardt, E.G., Goodman, B.N., Boyce, J.I., Lopez, G., van Hengstum, P., Rink, W.J., Mart, Y., and Raban, A. (2006) The tsunami of 13 December AD 115 and the destruction of Herod the Great's harbor at Caesarea Maritima, Israel. *Geology*, 34, 1061–1064.

Shtienberg, G., Yasur-Landau, A., Norris, R.D., Lazar, M., Rittenour, T.M., Tamberino, A., Gadol, O., Cantu, K., Arkin-Shalev, E., Ward, S.N., and Levy, T.E. (2020) A Neolithic mega-tsunami event in the eastern Mediterranean: Prehistoric settlement vulnerability along the Carmel coast, Israel. *Plos One*, 15, e0243619.

Stukeley, W. (1750) *The Philosophy of Earthquakes, Natural and Religious, or, an Inquiry into Their Cause, and Their Purpose* (Vol. 1). C. Corbet Publishers, London.

Tappin, D.R., Watts, P., McMurtry, G.M., Lafoy, Y., and Matsumoto, T. (2001) The Sissano, Papua New Guinea tsunami of July 1998—offshore evidence on the source mechanism. *Marine Geology*, 175, 1–23.

Chapter 8. Strand 7: Piled Higher and Deeper

Altinok, Y., Alpar, S.B., Ozer, N., and Vardar, H. (2011) Revision of the tsunami catalogue affecting Turkish coasts and surrounding regions. *Natural Hazards and Earth System Sciences*, 11, 273–291.

Bedford, S., Spriggs, M., and Regenvanu, R. (2006) The Teouma Lapita site and the early human settlement of the Pacific Islands. *Antiquity*, 80(310), 812–828.

Bedford, S., Spriggs, M., Buckley, H., Valentin, F., and Regenvanu, R. (2009) The Teouma Lapita site, South Efate, Vanuatu: A summary of three field seasons (2004–2006). In, Sheppard, P., Thomas T. and Summerhayes, G. (eds.) *Lapita: Ancestors and descendants*. Auckland: New Zealand Archaeological Association Monograph Series, 215–234.

Cain, G., Goff, J., and McFadgen, B.G. (2019) Prehistoric mass burials: Did death come in waves? *Journal of Archaeological Method and Theory*, 26, 714–754.

Chamberlin, T.C. (1890) The method of multiple working hypotheses. *Science*, 15, 92–96.

Goff, J., Charley, D., Haruel, C., & Bonte-Grapentin, M. (2008). Preliminary findings of the geological evidence and oral history of tsunamis in Vanuato. SOPAC Technical Report No. 416, 49.

Dawson, A.G., Dawson, S., and Bondevik, S. (2006) A late Holocene tsunami at Basta Voe, Yell, Shetland Isles. *Scottish Geographical Journal*, 122, 100–108.

Gillmore, G.K., and Melton, N. (2011) Early Neolithic sands at West Voe, Shetland Islands: Implications for human settlement. *Geological Society London, Special Publications*, 352, 69–83.

Goff, J., Chagué-Goff, C., Dominey-Howes, D., McAdoo, B., Cronin, S., Bonté-Grapetin, M., Nichol, S., Horrocks, M., Cisternas, M., Lamarche, G., Pelletier, B., Jaffe, B., and Dudley, W. (2011) Palaeotsunamis in the Pacific Islands. *Earth-Science Reviews*, 107, 141–146.

Goff, J., Goto, K., Ebina, Y., and Terry, J. (2016) Defining tsunamis: Yoda strikes back? *Earth-Science Reviews*, 159, 271–274.

Haflidason, H., Lien, R., Sejrup, H.P., Forsberg, C.F., and Bryn, P. (2005) The dating and morphometry of the Storegga Slide. *Marine and Petroleum Geology*, 22, 123–136.

Nitta Yoshisada. https://peoplepill.com/people/nitta-yoshisada. Accessed August 29, 2021.

Jones, W.H.S. (1918) *Pausanias: Description of Greece, Volume I, Books I and II (Attica, Corinth)*. Loeb Classical Library. William Heinemann; GP Putnam's Sons, New York.

Kritzas, Ch.B. (1976–1978) My kenaiko pegadi me skeletous sto Argos. Peloponnesiaka Parartema 6. *Proceedings of the first international colloquium on Peloponnesian*, Sparta, 7–14. 2: 173–180.

Leach, B., and Davidson, J. (2008) The Archaeology of Taumako. *New Zealand Journal of Archaeology Special Publication*, Wellington.

Little, L.M., & Papadopoulos, J.K. (1998) A social outcast in early Iron Age Athens. *Hesperia: The Journal of the American School of Classical Studies at Athens*, 67, 375–404.

Minami, M., Nakamura, T., Nagaoka, T., and Hirata, K. (2012) 14 C dating human skeletons from medieval archaelogical sites in Kamakura, Japan: Were they victims of Nitta Yoshisada's attack? *Radiocarbon*, 54, 599–613.

Saez, A., Margalef, O., Becerril, L., Herrera, C., Goff, J., Pla-Rabes, S., Becerril, L., Lara, L.E., and Giralt, S. (2022) Geological and climatic features, processes and interplay determining the human occupation of Easter Island. In, Rull, V., and Stevenson, C. (eds.) *The prehistory of Rapa Nui (Easter Island): Towards a multidisciplinary integrative framework. Developments in Paleoenvironmental Research, 22.* Springer, Cham, 311–344.

Smith, D.E., Shi, S., Cullingford, R.A., Dawson, A.G., Dawson, S., Firth, C.R., Foster, I.D., Fretwell, P.T., Haggart, B.A., Holloway, L.K., and Long, D. (2004) The Holocene Storegga slide tsunami in the United Kingdom. *Quaternary Science Reviews*, 23, 2291–2321.

Triantaphyllou, S., and Bessios, M. (2005) A mass burial at fourth century BC Pydna, Macedonia, Greece: Evidence for slavery? *Antiquity*, 79(305).

Walsh, S.L., Knüsel, C.J., and Melton, N.D. (2012) A re-appraisal of the Early Neolithic human remains excavated at Sumburgh, Shetland in 1977. *Proceedings of the Society of Antiquaries of Scotland*, 141, 3–17.

Chapter 9. Strand 8: A Country with Latitude

Aufderheide, A.C., Muñoz, I., and Arriaza, B. (1993) Seven Chinchorro mummies and the prehistory of northern Chile. *American Journal of Physical Anthropology*, 91, 189–201.

Castro, V., Aldunate, C., and Varela, V. (2012) Paisajes culturales de Cobija, costa de Antofagasta, Chile. *Revista chilena de antropología*, 26, 97–128.

Castro, V., Sanz, N., Arriaza, B.T., and Standen, V.G. (2014) Pre-Hispanic cultures in the Atacama Desert: A Pacific coast overview. *The Chinchorro Culture: A Comparative Perspective. The archaeology of the earliest human mummification*, UNESCO, Mexico City 11–34.

Cortés Olivares, H.F. (2005) El origen, producción y comercio del Pisco Chileno, 1546–1931. *Universum (Talca)*, 20, 42–81.

Dickinson, W.R. (1971) Plate tectonics in geologic history. *Science*, 174(4005), 107–113.

Goff, J., Chagué-Goff, C., Archer, M., Dominey-Howes, D., and Turney, C. (2012) The Eltanin asteroid impact—possible South Pacific palaeotmegatsunami footprint and potential implications for the Pliocene-Pleistocene transition. *Journal of Quaternary Science*, 27, 660–670.

Goff, J., Witter, R., Terry, J., and Spiske, M. (2020) Palaeotsunamis in the Sino-Pacific region. *Earth-Science Reviews*, 210, 103352.

Isbell, W.H. (2008) Wari and Tiwanaku: International identities in the central Andean Middle Horizon. In, Silverman, H., and Isbell, W. (eds.) *The handbook of South American archaeology*. Springer, New York, 731–759.

Kronmüller, E., Atallah, D.G., Gutiérrez, I., Guerrero, P. and Gedda, M. (2017) Exploring indigenous perspectives of an environmental disaster: Culture and place as interrelated resources for remembrance of the 1960 mega-earthquake in Chile. *International Journal of Disaster Risk Reduction*, 23, 238–247.

Le Roux, J.P. (2015) A critical examination of evidence used to re-interpret the Hornitos mega-breccia as a mass-flow deposit caused by cliff failure. *Andean Geology*, 42, 139–145.

Lomnitz, C. (2004) Major earthquakes of Chile: A historical survey, 1535–1960. *Seismological Research Letters*, 75, 368–378.

Rebolledo, S., Béarez, P., Salazar, D., and Fuentes, F. (2016) Maritime fishing during the Middle Holocene in the hyperarid coast of the Atacama Desert. *Quaternary International*, 391, 3–11.

Rice, P.M. (1996) Peru's colonial wine industry and its European background. *Antiquity*, 70, 785–800.

Rivera, M.A. (1991) The prehistory of northern Chile: A synthesis. *Journal of World Prehistory*, 5, 1–47.

Spiske, M., Bahlburg, H., and Weiss, R. (2014) Pliocene mass failure deposits mistaken as submarine tsunami backwash sediments—An example from Hornitos, northern Chile. *Sedimentary Geology*, 305, 69–82.

Vargas, G., Rutllant, J., and Ortlieb, L. (2006) ENSO tropical–extratropical climate teleconnections and mechanisms for Holocene debris flows along the hyperarid coast of western South America (17–24 S). *Earth and Planetary Science Letters*, 249, 467–483.

Victor, P., Sobiesiak, M., Glodny, J., Nielsen, S.N., and Oncken, O. (2011) Long-term persistence of subduction earthquake segment boundaries: Evidence from Mejillones Peninsula, northern Chile. *Journal of Geophysical Research: Solid Earth*, 116(B02402), 1–22.

Walsh, S.A., and Hume, J.P. (2001) A new Neogene marine avian assemblage from north-central Chile. *Journal of Vertebrate Paleontology*, 21, 484–491.

Walsh, S.A., and Martill, D.M. (2006) A possible earthquake-triggered mega-boulder slide in a Chilean Mio-Pliocene marine sequence: Evidence for rapid uplift and bonebed genesis. *Journal of the Geological Society*, 163, 697–705.

Weiss, R., Lynett, P., and Wünnemann, K. (2015) The Eltanin impact and its tsunami along the coast of South America: Insights for potential deposits. *Earth and Planetary Science Letters*, 409, 175–181.

Chapter 10. Strand 9: Life and Death on the Edge

Cocilovo, J.A., Varela, H.H., Quevedo, S., Standen, V., and Costa-Junqueira, M.A. (2004) La diferenciación geográfica de la población humana arcaica de la costa norte de Chile (5000-3000 AP) A partir del análisis estadístico de rasgos métricos y no métricos del cráneo. *Revista Chilena de Historia Natural*, 77, 679–693.

Goff, J., Witter, R., Terry, J., and Spiske, M. (2020) Palaeotsunamis in the Sino-Pacific region. *Earth-Science Reviews*, 210, 103352.

León, T, Vargas, G., Salazar, D., Goff, J., Guendón, J.L., Andrade, P., and Alvarez, G. (2019) Recording large Holocene paleotsunamis along the hyperarid coastal Atacama Desert in the major Northern Chile Seismic Gap. *Quaternary Science Reviews*, 220, 335–358.

Rebolledo, S., Béarez, P., Salazar, D., and Fuentes, F. (2016) Maritime fishing during the Middle Holocene in the hyperarid coast of the Atacama Desert. *Quaternary International*, 391, 3–11.

Salazar, D., Arenas, C., Andrade, P., Olguín, L., Torres, J., Flores, C., Vargas, G., Rebolledo, S., Borie, C., Sandoval, C., and Silva, C. (2018) From the use of space to territorialisation during the Early Holocene in Taltal, coastal Atacama Desert, Chile. *Quaternary International*, 473, 225–241.

Salazar, D., Easton, G., Goff, J., Guendon, J.L., González-Alfaro, J., Andrade, P., Villagrán, X., Fuentes, M., León, T., Abad, M., Izquierdo, T., Power, X., Sitzia, L., Álvarez, G., Villalobos, A., Olguin, L., Yrarrázaval, S., González, G., Flores, C., Borie, C., Castro, V., and Campos, J. (2022) Did a 3,800 years old ~Mw9.5 earthquake trigger major social disruption in the Atacama Desert? *Science Advances*, 8, eabm2996. https://www.science.org/doi/10.1126/sciadv.abm2996.

Thiel, M., Macaya, E.C., Acuña, E., Arntz, W.E., Bastias, H., Brokordt, K., Camus, P.A., Castilla, J.C., Castro, L.R., Cortés, M., Dumont, C.P., Escribano, R., Fernandez, M., Gajardo, J.A., Gaymer, C.F., Gomez, I., González, A.E., González, H.E., Haye, P.A., Illanes, J.-E., Iriarte, J.L., Lancellotti, D.A., Luna-Jorquera, G., Luxoro, C., Manriquez, P.H., Marín, V., Muñoz, P., Navarrete, S.A., Perez, E., Poulin, E., Sellanes, J., Sepúlveda, H.H., Stotz, W., Tala, F., Thomas, A., Vargas, C.A., Vasquez, J.A., and Vega, J.M. (2007) The Humboldt Current System of Northern and Central Chile: Oceanographic processes, ecological interactions and socioeconomic feedback. Ocean. *Oceanography and Marine Biology: An Annual Review*, 45, 195–344.

Chapter 11. Strand 10: The Future

Andrade, P., Castro, V., and Aldunate, C. (2016) Reconstrucción del modo de vida de individuos del arcaico de la costa arreica del norte de Chile: Una aproximación bioarqueológica desde el sitio Copaca 1. *Chungará (Arica)* 48, 73–90.

Andrade, P.A., Goff, J., Pearce, R.B., Cundy, A.B., Sear, D.A., and Castro, V. (2022) Evidence for a mid-Holocene drowning from the Atacama Desert coast of Chile? *Journal of Archaeological Sciences*, 140, 105565.

Chamberlin, T.C. (1890) The method of multiple working hypotheses. *Science*, 15, 92–96.

Evans, L. (2020) *Burying the dead: An archaeological history of burial grounds, graveyards and cemeteries*. Pen and Sword History, Barnsley, UK.

Olguín, L., Castro, V., Castro, P., Peña-Villalobos, I., Ruz, J., and Santander, B. (2015) Exploitation of faunal resources by marine hunter–gatherer groups during the Middle Holocene at the Copaca 1 site, Atacama Desert coast. *Quaternary International*, 373, 4–16.

Index

For the benefit of digital users, indexed terms that span two pages (e.g., 52–53) may, on occasion, appear on only one of those pages.

Figures are indicated by *f* following the page number